Birkhäuser

Lecture Notes in Geosystems Mathematics and Computing

About this series

The *Lecture Notes in Geosystems Mathematics and Computing* series showcases topics of current interest that lie at the interface of the geosciences, mathematics, and observational and computational sciences. Titles may present new results or novel perspectives on existing areas of research, or serve as extended surveys of certain topics. Traditional lecture notes covering core concepts for graduate and doctoral students are also included, as are titles that are based on expansions of outstanding PhD theses. Topically, titles address either mathematical and computational concepts and techniques – such as inverse problems, applied harmonic analysis, numerical simulation, and data analysis – or focus on practical applications in such areas as gravitation, climate modeling, seismology, geomechanics/dynamics, and satellite technology.

All manuscripts are peer-reviewed to meet the highest standards of scientific literature. Interested authors may submit proposals by email to the series editors or to the relevant Birkhäuser editor listed under “Contacts”.

Willi Freeden • Helga Nutz

Exploratory Potential Methods in Geothermal Power Generation

A Survey on Innovative Gravimetry and Magnetometry

Birkhäuser

Willi Freeden
Department of Mathematics, Geomathematics
University of Kaiserslautern-Landau
Kaiserslautern, Germany

Helga Nutz
CBM - Gesellschaft für Consulting
Business und Management mbH
Bexbach, Saarland, Germany

ISSN 2730-5996 ISSN 2512-3211 (electronic)
Lecture Notes in Geosystems Mathematics and Computing
ISBN 978-3-031-54411-8 ISBN 978-3-031-54412-5 (eBook)
https://doi.org/10.1007/978-3-031-54412-5

This book is published under the imprint Birkhäuser, www.birkhauser-science.com by the registered company Springer Nature Switzerland AG
The registered company address is: Gewerbestrasse 11, 6330 Cham, Switzerland

Paper in this product is recyclable.

Foreword

In the year 2006, the Ministry of Economics of Rhineland-Palatinate granted start-up funding for the establishment of a research team within the Geomathematics Group of the University of Kaiserlautern, so that a research platform, the so-called "Kaiserslautern Column Model (KCM)" (cf. Figure 0.1, see also [55] for more details) could be successfully developed and applied in the field of deep geothermal energy. KCM resulted in a total of eight Ph.-D. theses in the Geomathematics Group, some of which were published in book form by renowned publishing houses.

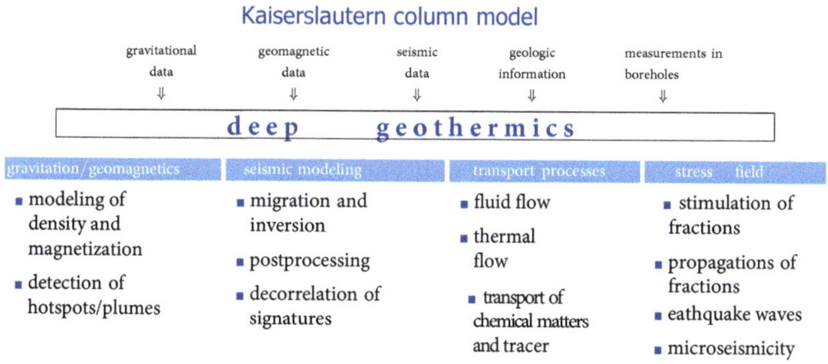

Fig. 0.1 Kaiserslautern Column Model (KCM) realized in the Geomathematics Group of the University of Kaiserslautern (2006-2016).

In the year 2012, essential parts of KCM went over to a research grant from the Federal Ministry of Economics Berlin within the framework of the joint project GEOFÜND (FKZ 0325512A-D), in which the LIAG Hannover, the Fraunhofer ITWM Kaiserslautern, the company G.E.O.S., Freiberg, and the Münchner Rückversicherung participated for a period of three years (PI Prof. Dr. W. Freeden, Geomathematics Group, University of Kaiserslautern). The main focus of the research activities was on exploratory seismic methods via new mollifier wavelet techniques, the identification of hot geothermal areas as well as the seismic specification of cost-effective drilling regions including the accompanying fractures in the deep underground.

In most European countries and, in particular, Germany, seismic is recognized as the standard method of exploration to such an extent that exploration is usually equated with seismic. However, in order to reduce the risk of drilling, it is often advisable to use gravimeter and/or magnetometer data for the analysis of geologic formations before extensive and cost-intensive seismic measurements are carried out. In order to make this observation more accessible for the exploratory practice and to provide a deepened justification, the company CBM - Gesellschaft für Consulting, Business und Management mbH, Bexbach, Saarland, decided to take up the ideas and concepts of the Geomathematics Group, University of Kaiserslautern, on decorrelative potential methods and to conduct intensive numerical test studies up to the point of usability in the exploration scenario. So, at the end of the year 2014, Prof. Dr. M. Bauer, CBM, and Prof. Dr. W. Freeden, University of Kaiserslautern, agreed to develop wavelet decorrelation of geologic structures, which had been successfully started using seismic as an example, also for the potential method of gravimetry under the auspices of the company CBM, and to validate its effectiveness in areas of the Saarland and its surroundings that are strongly influenced by anthropogenic mining. From May 2016 to April 2019, CBM realized the research project SPE (Satellite-Based Potential Methods for Geothermal Exploration, FKZ 0324061, PI Prof. Dr. W. Freeden, CBM manager Prof. Dr. M. Bauer) with the aim of developing novel concepts for the application of gravimetric potential methods for initial assessment in the planning and implementation of geothermal projects.

Based on the gravimetric research carried out in the Geomathematics Group, University of Kaiserslautern (2006-2016), the project SPE (2016-2019) provided the canonical understanding and consistent recognition that modern exploration first requires a broad range of geoscientific disciplines: In addition to geomathematics in its role as a key technology developing innovative decorrelation methods by means of multiscale analysis, geology, geophysics, and geoengineering are of essential importance. Geographers, geologists, geophysicists and geoengineers are responsible for data provision and processing as well as for the geologic evaluation and interpretation of innovative geomathematical modeling. In exploration, in fact, it is still considered a standard requirement for the successful application of potential methods based investigations that the geologic structures to be determined stand out from their surroundings by means of clear density contrasts, such as a salt dome or a magmatic hotspot. However, for example, the development of increasingly powerful absolute and relative gravimeters with significantly improved measuring accuracy and the applicability of new geomathematical mollifier methods within constructive approximation in recent years have shown that in future it will be certainly possible to detect weaker anomalies and model the "normal case of geologic layer sequences".

Since May 2019 up to April 2022, the company CBM and the TU Bergakademie Freiberg (Prof. Dr. C. Gerhards) in cooperation with the Interstate University of Applied Sciences Buchs NTB, Switzerland (Prof. Dr. M. Schreiner), had been implementing the BMWi (D) & BFO (CH) joint project "SYStematic EXPLoration" SYSEXPL (PI Prof. Dr. W. Freeden). The research project SYSEXPL was planned to run for a period of three years and was based on previous achievements of the projects GEOFÜND and SPE, in which new and efficient exploration methods for seismic and gravimetry were successfully investigated and tested. The numerical methods further developed in GEOFÜND and SPE from research results of the Geomathematics Group of the University of Kaiserslautern were transferred to magnetometry in the project SYSEXPL (PtJ FKZ 03EE4002A). The specific objective of the collaborative project SYSEXPL (PtJ FKZ 03EE4002A) was the research and development of innovative geomathematical decorrelation methods for application in magnetometry for geothermal exploration, a synoptic analysis of gravimetric decorrelation methods from the predecessor project SPE (PtJ FKZ 0324061) and the newly developed magnetometric decorrelation methods. These novel methods were prototypically applied to a test area Burbach of the Saarland using nonlinear approaches for a representation of the subsurface geology. Close cooperation took place with the project partner of the TU Bergakademie Freiberg (Prof. Dr. C. Gerhards) in the development of the analysis tool for inverse magnetometry from cleaned data. There was also a close exchange with the Swiss cooperation partner (Hochschule OST, research project REX, Prof. Dr. M. Schreiner) on the integration of the research results into the assessment tool for risk assessment of geothermal projects developed by the project REX. The final event "SYSEXPL: Systematic Exploration" at the IHK Saarland, Saarbrücken, on April 29, 2022, served to exchange information with experts and to make the project results known to the public as well as to experts.

Key literature on the development of decorrelative exploration methods that make up the survey content of the result report presented in this book are as follows: In an article published in "World of Mining" 2013, a novel method of regularizing data, the so-called mollifier methodology for decorrelation and inversion of exploration data, was proposed by W. Freeden, C. Blick [56], based on preliminary work by W. Freeden, M. Schreiner [77] in "Journal of Geodesy" (2006). Roughly speaking from a geothermal perspective, "mollification" entails the replacement of an "ill-posed" exploration problem by a system of well-posed problems that provide promising answers to ill-posedness and geologically evaluable proposed solutions. First attempts of mollifier exploration were documented numerically in the Ph.-D. thesis of the Geomathematics Group by C. Blick [28] for acoustic tomography. In the case of gravimetry, mollifier regularization using wavelets was described in detail for its mathematical/numerical ingredients by C. Blick, W. Freeden, H. Nutz [31], [32]. W. Freeden, M.Z. Nashed [64] gave a deep mathematical elaboration and methodological mollifier spline-wavelet advancements published in the "International Journal on Geomathematics" (2018). The monographs W. Freeden [54], C. Blick, W. Freeden, M.Z. Nashed, H. Nutz, M. Schreiner [34] provide an overview of the underlying theory so far and practice-relevant test and validation examples in gravimetry as well as magnetometry (additionally, C. Blick, S. Eberle [29], [30] lay the theoretical foundations for mollifier decorrelation methods in elastic seismics; the Ph.-D. thesis B. Kretz [113] deals with the transfer of essential components of mollifier methodology to poroelasticity). As a recent outcome (cf. [68]), a monograph by W. Freeden, M.Z. Nashed entitled "Recovery Methodologies–Regularization and Sampling" was published by AMS (American Mathematical Society) in August 2023, as a unifying contribution of ideas and concepts in mathematical recovery.

This work represents a numerical result report about topics to be found in the monographs W. Freeden [54], C. Blick, W. Freeden, M.Z. Nashed, H. Nutz, M. Schreiner [34], particularly in form of a georelevant collection of illustrative figures and tables. It pursues a double goal within the spectrum of geosciences: On the one side, it offers a scientific set of rules for today's geoengineering, which is interested in the application of innovative modeling and simulation techniques to promising datasets and structures. Special target groups are geoscientific institutions, engineering offices, public utilities and local energy providers as well as the entire exploration industry. On the other side, the book serves as a well understandable collection of current material in Applied Mathematics, Geophysics, and Geodesy.

In addition to the search for and acquisition of existing data, measurements of potential data of sufficiently high data density and accuracy had proven to be indispensable for purposes of decorrelative exploration. The measurement campaigns carried out by CBM, Bexbach, as the responsible institution, form an important building block for the provision of scientifically usable datasets and suitable geomathematical modeling and simulation. The programs developed in the projects SPE and SYSEXPL were designed in such a way that mainframes computers were not used as far as possible and office computers with appropriate dimensioning led to success.

Kaiserslautern, Bexbach, September 2023 *W. F., H. N.*

Preface

The book provides the geoscientific context, that arises in gravimetric/magnetometric exploration. The first step is to present a comprehensive formulation of the geoscientific framework. In addition, an insight into the current state of research is given by reducing gravimetry/magnetometry to mathematically accessible, thus, calculable and interpretable decorrelated models. In this way, the many unresolved questions and problems of gravimetry/magnetometry are to be made available to a broad geoscientific audience and the exploration industry in "mathematical appetizers". Innovative ways of utilization will be shown and new perspectives will be set in motion.

An essential characteristic of the book is its bridging function, in many ways. On the one side, the book leads in a loop from potential measurements by geoengineers particularly geodesists, through data cleansing by geophysicists, subsequent theory and model building, computer implementation and numerical calculation as well as simulation by geomathematicians, to interpretation by geologists, and back again if necessary. It thus spans the arc from geoengineering, especially geodesy, via geophysics to geomathematics and geology, and back again. On the other side, the book presents in clearly written condensed form the recently published pioneering and ground breaking multiscale mollifier methodologies realizing the transfer from the "reality space" of gravitational/magnetic measurements to the "virtual space" of mathematical/numerical mollifier wavelet decorrelations with novel geologic prospects as outcome.

The book essentially uses mathematics as a key technology for modeling and simulation issues on the basis of analysis and interpretation according to dense and precise gravitational/magnetic measurements. It is dedicated to surface and deep (volume) geology with potential data primarily of terrestrial origin. For deep geology, the geomathematical decorrelation methods are to be set up in such a way that borehole information may be canonically entered.

The work provides a consistent setup of concepts and ideas that have been already pursued by the Geomathematics Group, University of Kaiserslautern, over the last 15 years. The innovation of the work does not lie not in the basic theory of the individual chapters (in fact, some theoretical passages of the book have already been published elsewhere (see [34, 54]) in more detail according to their content). Instead, this work delivers the presentation of the underlying geomathematical philosophy and strategy in a unifying and clearly arranged synopsis, essentially on the basis of graphical illustrations and instructive numerical results.

An essential characteristic of the book is its bridging function, in many ways: The constituents of the bridging transfer from the "reality space" of observations to the "virtuality space" of modeling solutions and vice versa form the fundamental scientific background of the whole work. So, the book presents in clearly written condensed form the recently published pioneering and ground breaking multiscale mollifier methodologies realizing the bridging transfer from gravitational/magnetic measurements and observations to mathematical/numerical mollifier wavelet decorrelations of data with novel geologic prospects as outcome. Moreover, the book leads in a loop from potential measurements by geoengineers particularly geodesists, through data cleansing by geophysicists, subsequent theory and model building, computer implementation and numerical calculation as well as simulation by geomathematicians, to interpretation by geologists, and back again if necessary. The book spans the interdisciplinary arc from geoengineering, especially geodesy, via geophysics to geomathematics and geology, and back again. In more detail, the book presents in clearly written condensed form the recently published pioneering and ground breaking multiscale mollifier methodologies realizing the bridging transfer from gravitational/magnetic measurements to mathematical/numerical mollifier wavelet decorrelations with novel geologic prospects and results as outcome.

The book can be thought of as a results and validation presentation of the ideas and concepts initiated by [56] and developed in the works [54], [34]. Using the specific example of the German Saarland region, new important fields of application, especially for areas with mining-related cavities or densely built-up area in today's geoexploration, will be opened up and subjected to an in-depth geologic interpretation.

The success of the exploration techniques described in the book is based on the following circumstance: In explorative practice, it is still considered a standard prerequisite for the successful use of gravimetric/magnetometric investigations that the geologic structures to be determined are distinguished from their surroundings by clear density contrasts, such as a salt dome or an iron deposit. However, due to the development of increasingly powerful sensors with significantly improved measurement accuracy and the applicability of novel geomathematical methods in recent years, it appears that in the future it will also be possible to detect weaker anomalies and to model the "normal case of geologic layer sequences" by means of mollifier potential methods. Even more, synergetic effects can be obtained by comparison of gravimetric and magnetometric synopsis and the juxtoposition to seismic waves (if available).

Kaiserslautern, Bexbach, September 2023 *W. F., H. N.*

Acknowledgements

The authors thank the "Federal Ministry for Economic Affairs and Energy, Berlin" (today, "Federal Ministry for Economic Affairs and Climate Action, Berlin") and the "Projektträger Jülich" (PtJ) for funding the projects "GEOFÜND" (funding reference number FKZ 0325512A-D, PI Prof. Dr. W. Freeden, University of Kaiserslautern, Germany), "SPE" (funding reference number FKZ 0324061, PI Prof. Dr. W. Freeden, CBM - Gesellschaft für Consulting, Business und Management mbH, Bexbach, Germany, corporate manager Prof. Dr. M. Bauer), and "SYSEXPL" (funding reference number FKZ 03EE4002A&B, PI Prof. Dr. W. Freeden, CBM - Gesellschaft für Consulting, Business und Management mbH, Bexbach, Germany).

GEOFÜND (FKZ 0325512A-D), SPE (FKZ 0324061), and SYSEXPL (FKZ 03EE4002A&B) were funded by the German Federal Ministry of Economic Affairs and Energy within the framework of the 6th and 7th Energy Research Programme of the German Federal Government based on a resolution of the "Deutscher Bundestag", Berlin. The authors would like to express their sincere thanks to Dr. V. Monser and Dr. K. S. Schreiber, who were the responsible persons from PtJ and who accompanied the three research projects in an exemplarily excellent manner and always provided helpful advice and assistance. They thus made a decisive contribution to the success of decorrelative potential methods in exploration.

The authors would like to thank Prof. Dr. M. Bauer for the opportunity to continue the GEOFÜND-research activities of the Geomathematics Group, University of Kaiserslautern, within the framework of the company CBM - Gesellschaft für Consulting, Business und Management mbH, Bexbach, Saarland, and to kindly take over all administrative tasks related to the research projects SPE and SYSEXPL.

The SPE measurement campaign carried out by CBM was supported in a competent manner in technical terms by the state offices of the federal states of Saarland and Rhineland-Palatinate. Special thanks go to Dipl.-Ing. T. Lehnert and Dipl.-Ing. (FH) M. Müller (State Office for Surveying, Geoinformation and Land Development Saarland, Saarbrücken) as well as Dipl.-Ing.

G. Berg and Dipl.-Ing. (FH) M. Cieslack (State Office for Surveying and Basic Geographic Information Rhineland-Palatinate, Koblenz).

The following comments can be made about the staffing and structure of the projects: Prof. Dr. W. Freeden was the scientific leader of all three projects. Dr. Nutz was in charge of the geomathematical group, Dipl.-Ing. T. Neu was in charge of the non-mathematical group. In the execution of the projects, the compilation and presentation of results in publications and reports, the management was substantially supported by the staff members Dr. C. Blick (from September 2022 professor at the University of Applied Sciences Kempten), Dr. T. Degro, M. Sc. E. Byamba, Dipl.-Geogr. B. Jakobs, Dipl.-Geol. Z. Hauler, M. Sc. B. Kretz, M. Sc. D. Krüger and Dipl.-Geophys. L. Mailänder. Without the integrative cooperation of all participants, a data acquisition by measurement, an implementation of the proposed geomathematical concepts, and the completion of the illustration and application part of this book would not have been possible.

W. Freeden would like to thank his friend M. Zuhair Nashed, Mathematics Department, University of Central Florida, Orlando, USA, for numerous fruitful conversations and mathematical collaboration over the past 25 years.

Both authors of this work owe a great debt of gratitude to Birkhäuser, in particular T. Hempfling and K. Kambaso, for their willingness to accept the book in the series “Lecture Notes on Geosystems Mathematics and Computing”.

About the Authors

Willi Freeden, University of Kaiserslautern, Mathematics Department, Kaiserslautern, Rhineland-Palatinate, Germany

Studies in Mathematics, Geography, and Philosophy at the RWTH Aachen University, 1971 Diplom in Mathematics, 1972 Staatsexamen in Mathematics and Geography, 1975 Ph.-D. thesis in Mathematics (RWTH Aachen University), 1979 Habilitation in Mathematics (RWTH Aachen University),

1981/1982 Visiting Research Professor at The Ohio State University, Columbus (Department of Geodetic Science and Surveying),

1984 Professor of Mathematics at the RWTH Aachen University (Institute of Pure and Applied Mathematics),

1989 Professor of Technomathematics (Industrial Mathematics) at the University of Kaiserslautern,

1994 Head of the Geomathematics Group at the University of Kaiserslautern,

2002 - 2006 Vice-President for Research and Technology at the University of Kaiserslautern,

Organized more than 45 international conferences and mini-symposia all over the world (organizer of several Oberwolfach conferences, mini-symposia and Special Sessions at meetings of the American Mathematical Society (AMS)),

Published 25 books as author and/or editor and more than 225 journal contributions (in mathematics and geosciences),

Member of the editorial board of a large number of international journals,

2009 (founding and executive) Editor-in-Chief of the Springer "International Journal on Geomathematics (GEM)",

2010 Editor-in-Chief (together with T. Sonar, M.Z. Nashed) of the Springer "Handbook of Geomathematics", (First Edition), 2015 Springer "Handbook of Geomathematics" (Second Edition),

2014 Editor-in-Chief (together with M.Z. Nashed) of the Birkhäuser book series "Geosystems Mathematics",

2014 Editor-in-Chief (together with M.Z. Nashed, O. Scherzer) of the Birkhäuser "Lecture Notes on Geosystems Mathematics and Computing",

2014 Editor-in-Chief (together with M. Bauer, H. Jacobi, T. Neu) of the (German) Springer-Spektrum "Handbuch Tiefe Geothermie",

2017 Editor-in-Chief (together with R. Rummel) of the (German) Springer-Spektrum "Handbuch der Geodäsie",

2018 Editor-in-Chief (together with M. Bauer, H. Jacobi, T. Neu) of the (German) Springer-Spektrum "Handbuch Oberflächennahe Geothermie",

2018 Editor-in-Chief (together with M.Z. Nashed) of the Birkhäuser "Handbook of Mathematical Geodesy",

2020 Editor-in-Chief of the Springer-Spektrum "Handbuch Mathematische Geodäsie/Mathematical Geodesy",

2022 Editor-in-Chief (together with M.Z. Nashed) of the World Scientific Encyclopedia "Frontiers in Entropy Across the Disciplines",

1975 Borchers Medal, Award of the RWTH Aachen University,

1989 Fellow of the International Association of Geodesy (IAG),

1996 Member of the German Geodetic Commission of the Bavarian Academy of Sciences, Munich,

2018 IPMS (Inverse Problems: Modeling & Simulation) Award,

2020 Vening Meinesz Medal: Award of the European Geosciences Union (EGU).

Helga Nutz, CBM - Gesellschaft für Consulting, Business und Management mbH, Bexbach, Saarland, Germany

Studies in Technomathematics at the University of Kaiserslautern,

1998 Diploma in Technomathematics

2002 Ph.-D. thesis in Mathematics (University of Kaiserslautern, Geomathematics Group),

2002 - 2003 Research Officer at the University of Kaiserslautern,

2006 - 2016 Research associate at the Geomathematics Group (Head: Prof. Dr. W. Freeden) of the University of Kaiserslautern,

2012 - 2016 Coordinator in the BMWi joint project GEOFÜND,

2016 - 2022 Research associate at CBM - Gesellschaft für Consulting, Business und Management mbH, Bexbach, and there deputy project manager in the BMWi projects SPE and SYSEXPL.

About the Research Team

Mathias Bauer, Prof. Dr. habil., Dr. h.c.

Professor, RWTH Aachen University, occupational safety and health protection, head of the company "CBM - Gesellschaft für Consulting, Business und Management mbH, Bexbach, Saarland, Germany",
SPE/SYSEXPL corporate manager.

Christian Blick, Prof. Dr.

Professor, University of Applied Sciences, Kempten (Allgäu), Bavaria, Germany,
GEOFÜND/SPE/SYSEXPL geomathematician.

Erdenebataar Byamba, M.Sc.

CBM project engineer,
SPE/SYSEXPL surveyer and geodesist.

Thomas Degro, Dr.

CBM consultant for gravimetry,
SPE geophysician.

Willi Freeden, Prof., PD., Dr.

Professor, University of Kaiserslautern (today, University of Kaiserslautern-Landau), Department of Mathematics, GEOFÜND/SPE /SYSEXPL project leader
GEOFÜND/SPE/SYSEXPL geomathematician.

Zita Hauler, Dipl.-Geol.

CBM research associate,
SPE/SYSEXPL geologist.

Bernd Jakobs, Dipl.-Geograph.

CBM commissioner, research associate,
SPE/SYSEXPL geographer.

Laura Mailänder, M.Sc.

CBM research associate,
SPE/SYSEXPL geophysician.

Thomas Neu, Dipl.-Ing.

CBM research associate and SPE/SYSEXPL deputy project manager (measurement and data),
SPE/SYSEXPL geoengineer.

Helga Nutz, Dr. Dipl.-techn.

CBM research associate and SPE/SYSEXPL deputy project manager (modeling),
GEOFÜND/SPE/SYSEXPL geomathematician.

Contents

Chapter 1

Geothermal energy and exploration

Abstract:

Our interest in this introductory chapter is in mentioning that geothermal energy implies a secure and stable basic supply of electricity and heat. Besides the advantages (e. g., independence of season and weather phenomena, environmentally friendly character, optically unobtrusive and minor space requirement, optimal energy yield by power-heat-coupling) there is as critical point the reduction of the geothermal exploration risk, which overcoming needs enhanced research activities. The geoscientific basis is explained for the strong conviction that potential methods such as gravimetry and magnetometry are quite competitive exploration methods to seismic in the future.

1.1 Geothermal reservoirs

Geothermal energy is etmylogically composed of three words according to its Greek origin: $\gamma\tilde{\eta}$ =Erde, $\theta\epsilon\rho\mu\acute{o}\varsigma$ = warm, and $\acute{\epsilon}\nu\acute{\epsilon}\rho\gamma\epsilon\iota\alpha$ =Wirksamkeit. Nowadays, geothermal energy is an important component due to the increasing need of renewable energies production.

Temperature, in fact, increases with depth in the Earth at the average of 25 $°C/km$. If the average surface temperature is assumed to be around 15 $°C$, the temperature at about 3 km is around 90 $°C$. All in all, in accordance with the standard geophysical estimates, the total heat content of the Earth (reckoned at an average surface temperature of 15 $°C$) is approximately $12.6 \cdot 10^{24} MJ$, where the heat content of the crust amounts to $5.4 \cdot 10^{21} MJ$.

W. Freeden, H. Nutz, *Exploratory Potential Methods in Geothermal Power Generation*, Lecture Notes in Geosystems Mathematics and Computing,
https://doi.org/10.1007/978-3-031-54412-5_1

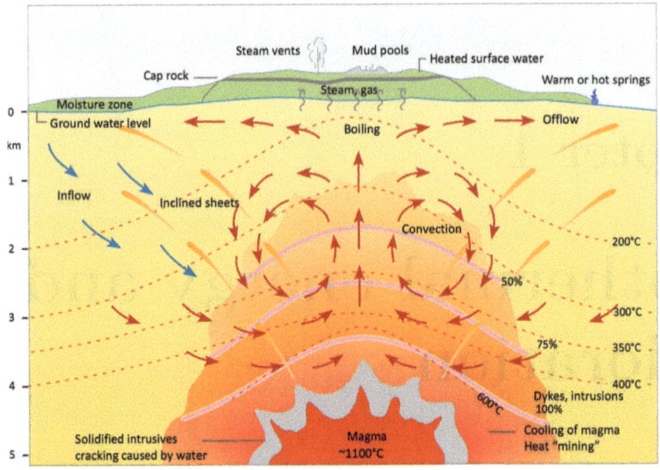

Fig. 1.1 Schematic representation of geothermally relevant zones [150].

Nowadays only a small fraction of the heat content is utilized, depending on the geologic knowledge. In favour are areas that transfer heat from deep zones to the surface. Spatial variations of thermal energy within the deep crust and mantle of the Earth give rise to concentrations of thermal energy near the surface of the Earth, such that certain locations can be used as an energy resource. Heat is transferred from the deeper parts of the Earth by conduction through rocks, by the movement of hot deep material towards the surface, particularly when associated with recent volcanism, and by circulation of water, e. g., in active fault zones. Figure 1.1 illustrates some scenarios of interest for geothermal exploitation.

Much of the geothermal exploration occurring worldwide is focused on the geologic concept of plate tectonics since most of the current thermal activities are located near plate boundaries. The brittle and moving plates of the lithosphere (crust and upper mantle) are driven by convection of plastic rocks beneath the lithosphere. Convection causes the crustal plates to break and move away in opposite directions from zones of upwelling hot material. Magma moving upward into a zone of separation is followed by substantial amounts of thermal energy. A particular source of elevated heat flow and volcanism are plumes and hotspots. Areas of the world with high geothermal potential are shown in Figure 1.2.

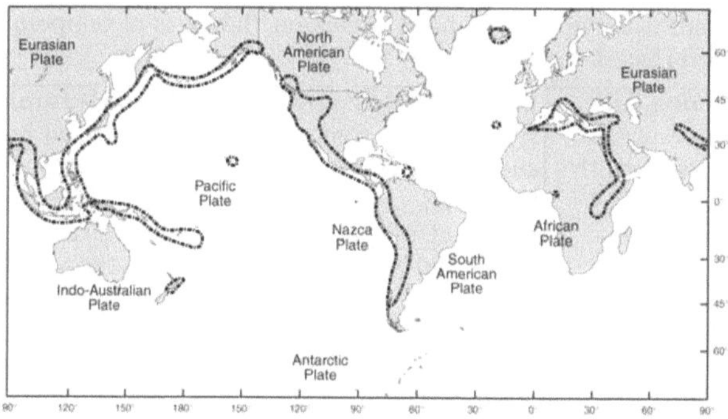

Fig. 1.2 Areas of the Earth with larger potential for generating geothermal energy (cf. [94]).

1.2 Geothermal system classification

Innovative techniques of exploration, geothermal drilling, electric power generation, heat pumps, etc. may open new frontiers in the near future and probably allow utilization of larger fractions of the untapped resources. In fact, the heat in place is not the problem. The question is how much can be taken out and at what price.

Indeed, the complexity of the entire energy production chain makes modeling of geothermal reservoirs a difficult challenge involving different scientific disciplines. For the geothermal energy production system to be efficient, it is of essential importance not only to have a deep understanding of the geologic, thermal, and mechanical configuration of the potential site, but also to be able to predict possible consequences of this invasive procedure, to satisfy mathematical requirements, and to execute large numerical computations. For this reason, geologists, engineers, physicists, and mathematicians must share their expertise and work together in order to provide an efficient solution for the ambitious task "geothermal energy".

The gained geothermal energy can be employed directly for the heat market or it can be used indirectly for electricity generation. Countries with active volcanism such as Iceland have a longstanding tradition of industrial use of geothermal energy gained by reservoirs with high enthalpy ($> 180\,^{\circ}$C), which are easily accessible due to their shallow depths (cf. [85]). Nevertheless, the absence of these sources does not exclude substantial geothermal potential in other regions or countries in the shape of deep reservoirs with low enthalpy ($< 180\,^{\circ}$C) occurring in sediment basins or in the vicinity of reservoirs with high enthalpy.

Table 1.1 Classification of geothermal systems on the basis of temperature, enthalpy, and physical state (based on the work by [12, 35, 154].

<table>
<tr><td>Low-temperature (LT) systems with reservoir temperature at 1km depth below 150°C. Often characterized by hot or boiling springs.</td><td rowspan="2">Low-enthalpy geothermal systems with reservoir fluid enthalpies less than 800kJ/kg, corresponding to temperatures less than about 180°C.</td><td rowspan="2">Liquid-dominated geothermal reservoirs with water temperature at, or below, the boiling point at the prevailing pressure, such that the water phase controls the pressure in the reservoir. Some steam may be present.</td></tr>
<tr><td>Medium-temperature (MT) systems with reservoir temperature at 1km depth between 150 – 200°C.</td></tr>
<tr><td rowspan="2">High-temperature (HT) systems with reservoir temperature at 1km depth above 200°C. Characterized by fumaroles, steam vents, mud pools and highly altered ground.</td><td rowspan="2">High-enthalpy geothermal systems with reservoir fluid enthalpies greater than 800kJ/kg.</td><td>Two-phase geothermal reservoirs where steam and water co-exist and the temperature and pressure follow the boiling point curve.</td></tr>
<tr><td>Vapour-dominated geothermal systems where temperature is at, or above, the boiling point at the prevailing pressure and the steam phase controls the pressure in the reservoir. Some liquid water may be present.</td></tr>
</table>

Following the classification presented in Table 1.1, all geothermal reservoirs can be roughly grouped into three categories based on their nature and geologic settings:

(1) Volcanic geothermal systems are in one way or another associated with volcanic activity. The heat sources for such systems are hot intrusions or magma. They are most often situated inside, or close to, volcanic complexes at or near plate boundaries or hotspots. Permeable fractures and fault zones but also permeable strata such as sediments, ignimbrite, or lava control the flow of water in volcanic geothermal systems.
(2) In convective systems, the heat source is the hot crust at depth in tectonically active areas, preferably with above average heat flow. Here, the geothermal water has circulated to considerable depth ($> 1\ km$), through mostly vertical fractures, to extract the heat from the rocks.
(3) Sedimentary geothermal systems are probably the most common type worldwide. The systems in this category are characterized by medium temperature, high flow rate geothermal reservoirs in large-basin, sedimentary deposits.

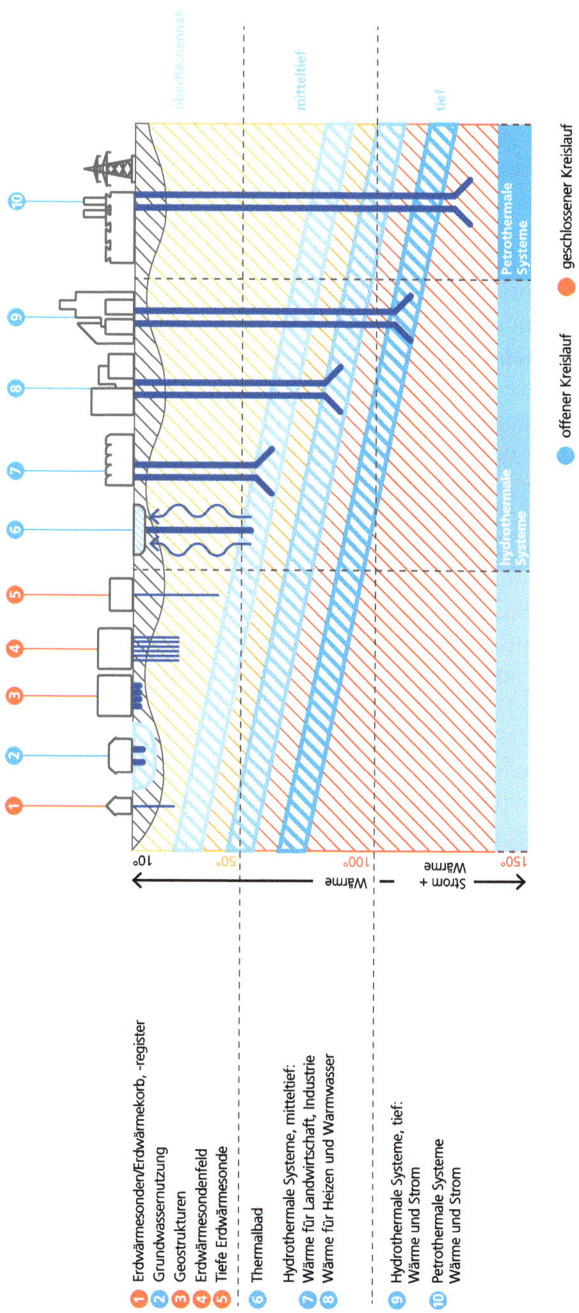

Fig. 1.3 Diversity of production facilities for obtaining geothermal energy (in german, Bundesamt für Energie, Schweiz, 2017), see also [59].

Geothermics can also be classified based on depth. It distinguishes between near-surface geothermal energy and deep geothermal energy (cf. [22, 23]. The former is mainly used for heating in private homes. The required facilities are, e. g., geothermal collectors or thermo-active pipes reaching down to a depth of at most 400 m. The use of deep geothermal energy, on the other hand, requires boreholes with a depth between 2 and 5 km.

Basically, there are three different types of systems for deep geothermal power production (see Figure 1.3): thermowells (deep heat exchanger), hydrothermal reservoirs, and petrothermal reservoirs. In a thermowell, the heat transfer medium circulates in a closed cycle within a U-pipe or a coaxial heat exchanger, therefore only one borehole has to be drilled. Although there is the advantage of no contact with the groundwater, the relatively low productivity gives rise to focus on open systems, namely hydrothermal and petrothermal reservoirs. The concept behind hydrothermal systems is to let thermal water found in deep reservoirs circulate between two or three drilled deep wells through a previously existing aquifer. Typically, these reservoirs consist of a porous medium layer heated from below by a hot stratum of impermeable material. By contrast, in petrothermal systems (also called Hot-Dry-Rock Systems or Enhanced Geothermal Systems (EGS)) the water flows through fractured hot rock, the porosity of which can be enhanced by hydraulic stimulation. In the latter case, water is artificially pumped into the reservoir.

It is our belief that a geoscientific consortium including geomathematics as a key technology can contribute significantly to find a more appropriate answer for all types of geothermal energy systems, i. e., deep as well as near surface geothermal systems (see Figure 1.3). The key issues that geomathematicians have to face are the detection of potential reservoirs along with the surrounding subsurface structure including temperature, capacity and hydraulic characteristics of aquifers and the development of a comprehensive model describing the dynamics of the production process, in particular as far as flow, temperature, and composition of the fluid are concerned. Recent events such as earthquakes (see, e. g., [146]) show that another fundamental mathematical problem is the understanding of inner stress conformation and dynamics.

1.3 Geothermal energy as renewable energy

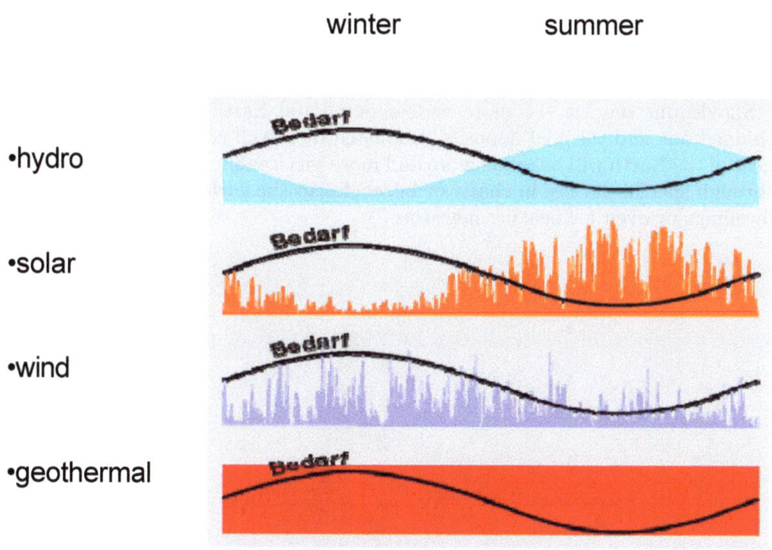

Fig. 1.4 Demand ("Bedarf") and availability of renewable energy in Germany (from [59]). The graphs show the exceptional position of "geothermal energy" as renewable energy beeing constantly available throughout the year.

Renewable energies or regenerative energies are energy sources that are inexhaustibly available to mankind or are renewed relatively quickly. This distinguishes them from fossil fuels. Renewable energy sources include bioenergy, hydropower, marine energy, solar energy, wind energy, and geothermal energy (geothermal heat). Today, the rapid realization of the ambitious goals of decarbonizing of energy production and overcoming shortages of energy sources, especially in times of war, can only be achieved if, in addition to sun, water and wind, geothermal energy (see Figure 1.4), characterized by its constant availability and year-round usability, is adequately appreciated.

As for geothermal energy, only one percent of our Earth's body is less hot than 1000 $°C$. Of this one percent of the Earth's heat, again only one percent is less hot than 100 $°C$. As a result, geothermal energy is one of the promising renewable energy sources. Energy production by means of geothermal power stations taps the heat stored in the accessible part of the Earth's crust.

The need to develop geothermal energy reservoirs from the Earth was formulated as early as 1858:

> "Sollten einst auf der mehr oder weniger bevölkerten Erde die Wälder so stark gelichtet und die Kohlenlager erschöpft sein, so ist es wohl denkbar, dass man die Innenwärme der Erde sich mehr und mehr dienstbar macht, dass man sie durch besondere Vorrichtungen in Schächten oder Bohrlöchern zur Oberfläche leitet und zur Erwärmung der Wohnungen oder selbst zur Heizung von Maschinen verwendet" (Carl Bernhard von Cotta, Freiberger Bergrat und Mineraloge, cf. Figure 1.5).
>
> ("Should one day on the more or less populated Earth the forests be so much thinned out and the coal deposits exhausted, it is well conceivable that the inner heat of the Earth will be made more and more serviceable, that it will be conducted through special devices in shafts or boreholes to the surface and used for heating dwellings or even for heating machines.")

Fig. 1.5 Carl Bernhard von Cotta (1808 - 1879).

However, it was not until the 20th century that the use of geothermal energy, which is dictated by nature and thus tied to the location, became possible. The first modern district heating system based on geothermal energy was installed in Reykjavik (Iceland). Networks with geothermal energy in France, Italy, Hungary, Romania, Turkey, Georgia, China were started during the 1970s, mainly as a result of the oil crises. In the vicinity of Paris and in the French capital itself, a good 250 000 homes have been geothermally heated since the 1970s.

The need for renewable energy is growing every day - it is increasingly determining today's economic situation and development. This is caused, among other things, by the increased environmental awareness due to climate change and by the increasing scarcity of economically and ecologically justifiable extractable fossil raw materials. Nonetheless, in Germany, the use of geothermal energy is largely unknown or even rejected by the public. Moreover, there often are obstacles by politicians to accept geothermal energy as a proper form of renewable energy. The resulting lack of attention, however, bears no relation to the achievable possibilities offered by this regenerative form of energy: In fact, the heat content of the Earth would cover our current world energy needs for 30 million years. By human standards, then, the energy reserves stored in the Earth are "just as inexhaustible" as those of the sun.

In the Earth's crust, the temperature increases on average by about 3 $°C$ per 100 m ("geothermal gradient"). However, it must be emphasized that the geothermal heat is not evenly distributed. A fundamental prerequisite for the safe implementation of geothermal projects is the most accurate possible knowledge of the situation in the subsurface for the area of the planned drilling. This can significantly reduce both the risk of drilling the wrong borehole and the risk of Earth movements after the geothermal power station is put into operation. Geothermal power stations can indeed trigger Earth movements in the subsurface. As a rule, however, these are far below the threshold of perception and damage (microseismicity).

1.4 Geothermal energy and reflection seismics

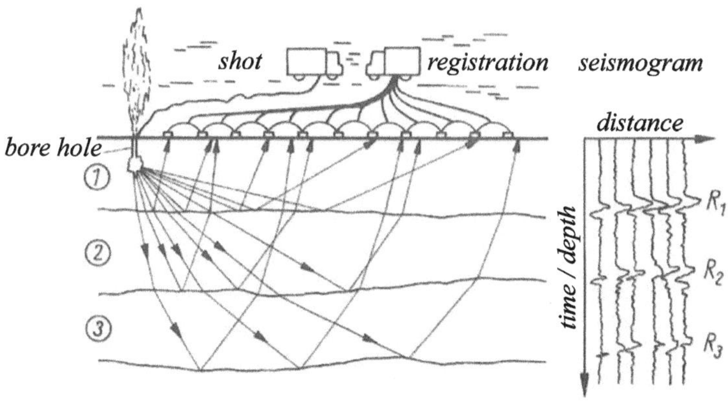

Fig. 1.6 Seismic acquisition (modified illustration taken from [104] with kind permission of the publisher).

Seismic exploration is the most commonly practiced exploration technique in Germany. Seismic measures the travel time of wave or pulse-shaped signals passing through the geologic formations. If the speed of signal propagation is known (cf. Figure 1.6), conclusions can be drawn about the distance traveled and thus about the internal structure. An illustration of reflection seismicity to determine the migration of a transmitted wave signal is documented graphically in time sequence in the Figure 1.9 (cf. [102]). More explicitly, Figure 1.9 shows the wave propagation specifically in the velocity field of the Marmousi dataset, which often serves as a geologic reference in seismic data processing. The so-called Marmousi model (cf. Figure 1.7) was generated in 1988 by the "Geophysics Group in the Institut Francais du Petrole" and is based on a profile through the "North Quenguela Trough in the Cuanza Basin in Angola". The model is discretely available in a grid of 384×122 sample points of a spacing of 24 m. Thus, it measures 9192 m in length (x_1 direction) and 2904 m in depth (x_3 direction). For more information about the Marmousi model, we refer to the extensive literature, in particular [180].

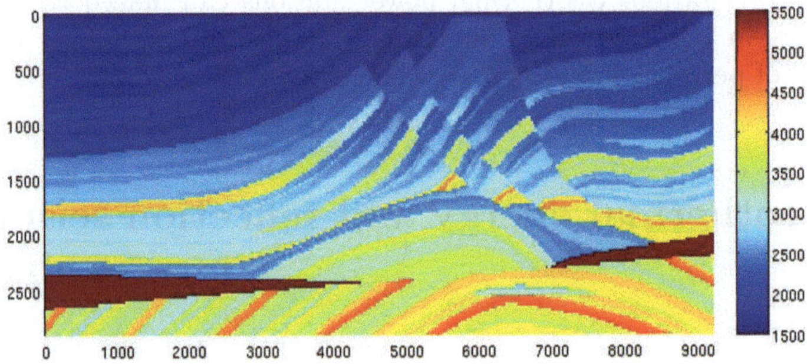

Fig. 1.7 Marmousi velocity field in $[\frac{m}{s}]$.

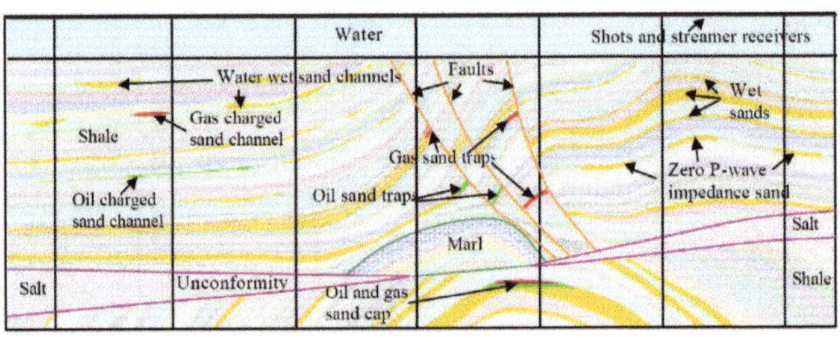

Fig. 1.8 Marmousi interpretation: lithology and model features including geologic formation boundaries.

Wave propagation in the Marmousi velocity model after 0.6 s

Wave propagation in the Marmousi velocity model after 1.1 s

Wave propagation in the Marmousi velocity model after 1.6 s

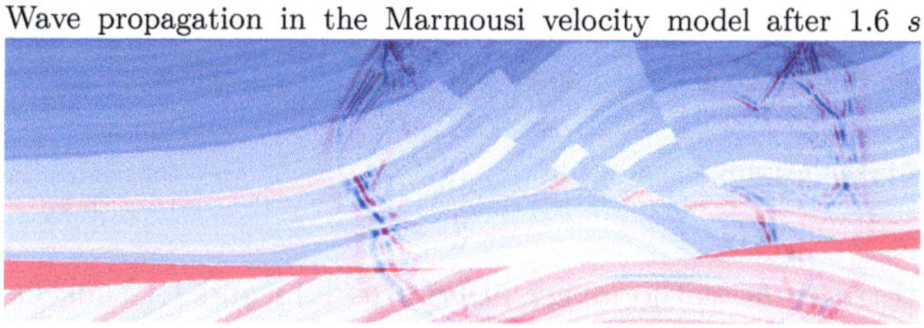

Wave propagation in the Marmousi velocity model after 2.1 s

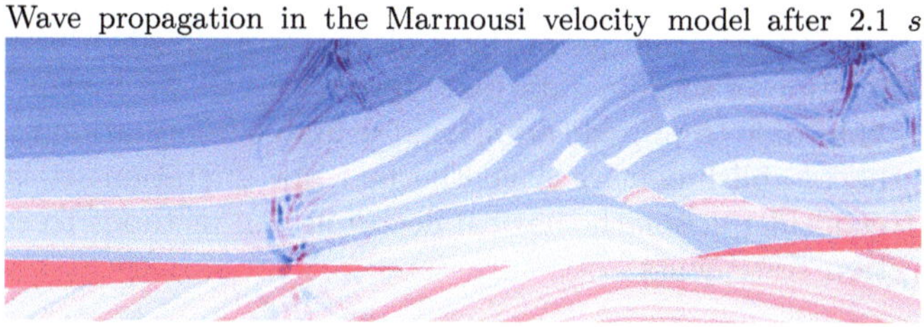

Wave propagation in the Marmousi velocity model after 2.6 s

Fig. 1.9 Visualization of wave propagation in the Marmousi velocity field (taken from the Ph.-D. thesis [102]).

The geologic interpretation of the Marmousi model (shown in Figure 1.7) provides Figure 1.8 according to [120]. Striking are the faults in the layering caused by an underlying plume.

Based on the velocity field (see Figure 1.7), a migration result (see Figure 1.10) was generated for testing purposes in the Ph.-D. thesis M. Ilyasov [102]. In the Ph.-D. thesis C. Blick [28] the potential of the scattered wavefield was generated. Both were subsequently subjected to a wavelet decorrelation mechanism in the GEOFÜND project (PtJ FKZ 0325512A) (cf. Figure 1.11 for the velocity field and Figure 1.12 for the migration field) (more details and illustrations can also be found in [53, 55]).

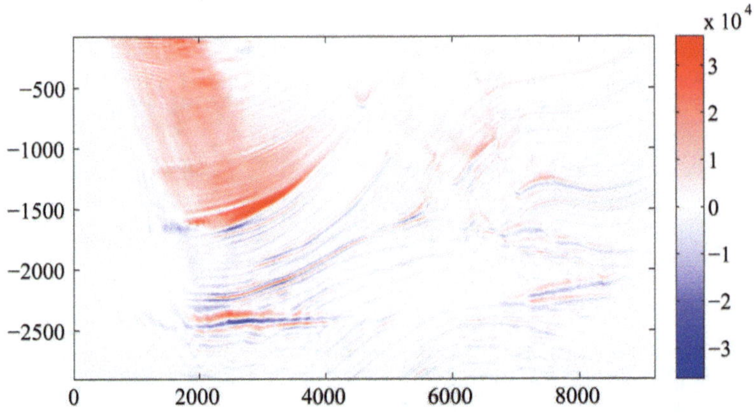

Fig. 1.10 Marmousi migration result.

The decorrelation methodology in Figures 1.11 and 1.12 successfully provides the transitions of the underlying lithology, which with increasing wavelet filter scale even show an increasingly concise structure up to the detection of faults (discontinuities of the geologic formations). A wavelet (mollifier) decorrelation of seismic data (as impressively shown by the Marmousi example) is thus able to provide additional valuable information for risk mitigation of the establishment of a geothermal power station. It is therefore advisable to consider geomathematical decorrelation methods to ensure seismic exploration statements.

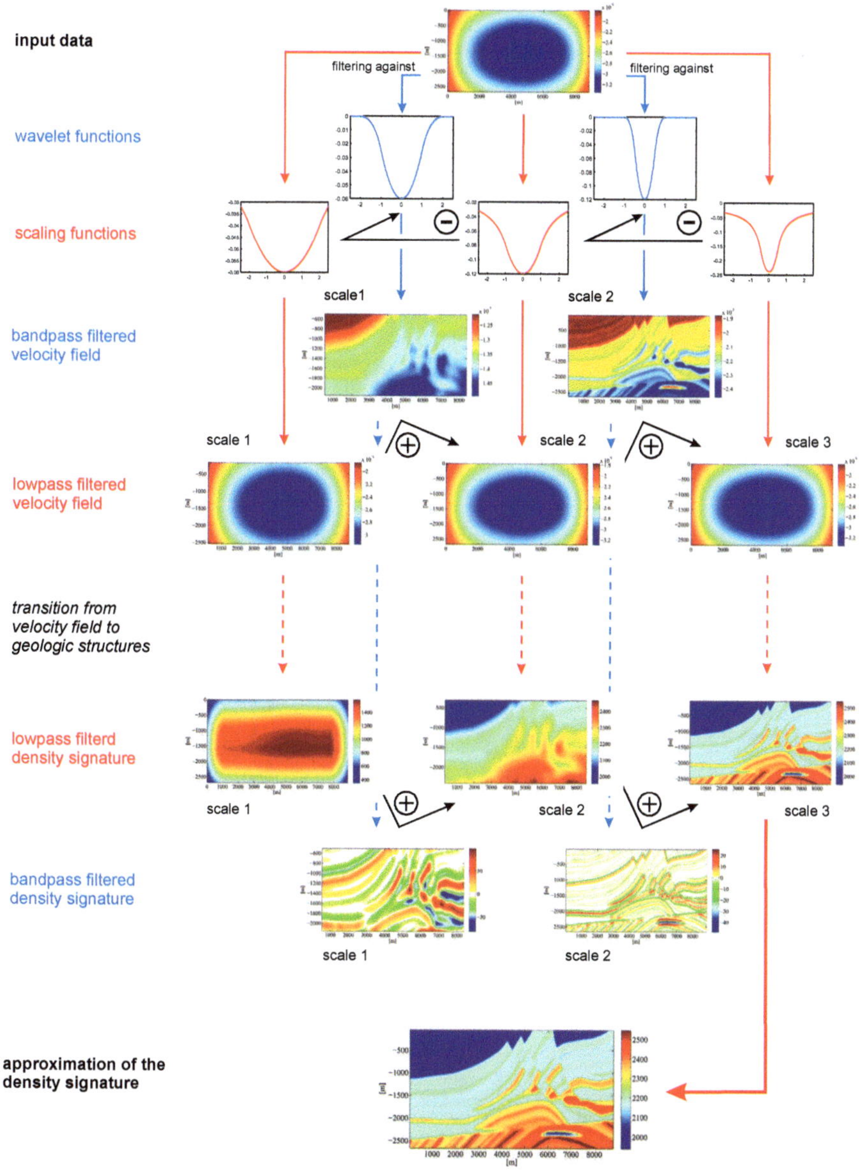

Fig. 1.11 Schematic visualization of the methodology of decorrelation of the Marmousi velocity field and transition to density (GEOFÜND project (PtJ FKZ 0325512A)).

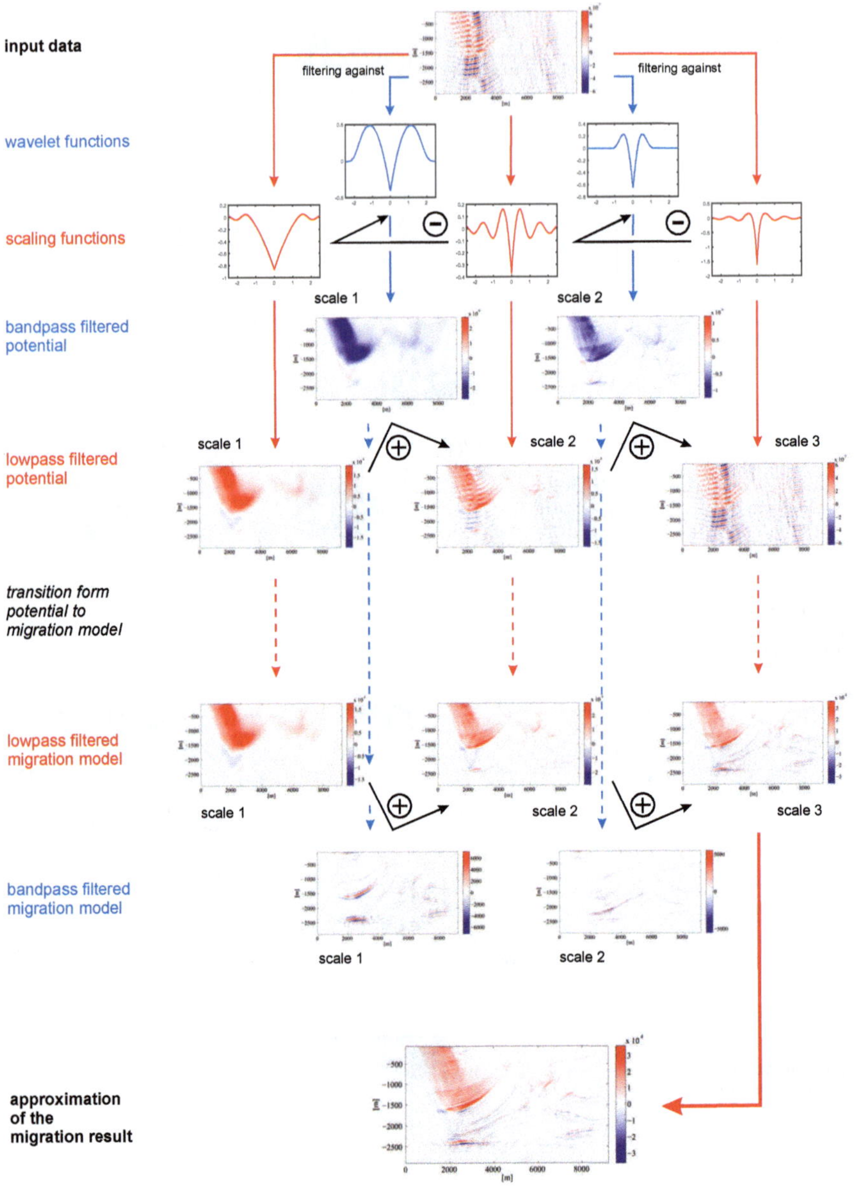

Fig. 1.12 Schematic visualization of the scattered wave potential methodology and the decorrelation transition to the migration model (GEOFÜND project (PtJ FKZ 0325512A)).

As a résumé we are confronted with the following situation: In the run-up to a geothermal well, as far as the planning and construction of a geothermal power station is concerned, a target area for the intended well must be identified in the course of a careful prospecting and exploration usually first based on the evaluation of data accessible free of charge. In the further course, based on purchased or commissioned 2D and 3D seismics as well as a synthesis of all geothermal-geologic data, a determination of the drilling targets as well as the determination of the drilling paths has been carried out in the course of the exploration.

Furthermore it should be noted that in the context of reflection seismicity, waves are triggered by artificial shaking, which, according to the low energy, penetrate only the uppermost part of the Earth's crust. Phenomena of microseismicity are thus unpreventable for reflection seismics. Therefore, the use of seismic often meets with strong rejection in the German population. As a consequence, acceptance problems arise with regard to seismics and, in a completely unacceptable way, also with regard to geothermal energy per se.

Even if the subsurface is suitable for performing a 3D seismic, two other aspects must be critically seen:

- The execution of a 3D seismic survey by using heavy vibration vehicles corresponding traffic obstruction is always associated with damage to roads, supply and disposal lines and occasionally also to houses, the removal or compensation of which often leads to discussions with the owners. This has a further negative impact on the public image and continues to cause discussion for years to come.

- Planning, approval and implementation of a 3D seismic survey cannot be achieved in less than two years, since for reasons of flora and fauna protection, measurements can usually only be made in the winter half year (for further details the reader is referred to a DMT timeline representation for the implementation of a deep geothermal project by S. Bißmann, "Einmal Seismik bitte – aber was genau ist richtig?", presentation at the DMT seismic webinar, 02-17-2022). More recent considerations also aim at the summer half year, when extensive environmental protection tests are carried out beforehand. Even if it seems possible from a licensing point of view to halve this period, many investors find it takes too long until the first detailed results relevant for decision-making are available.

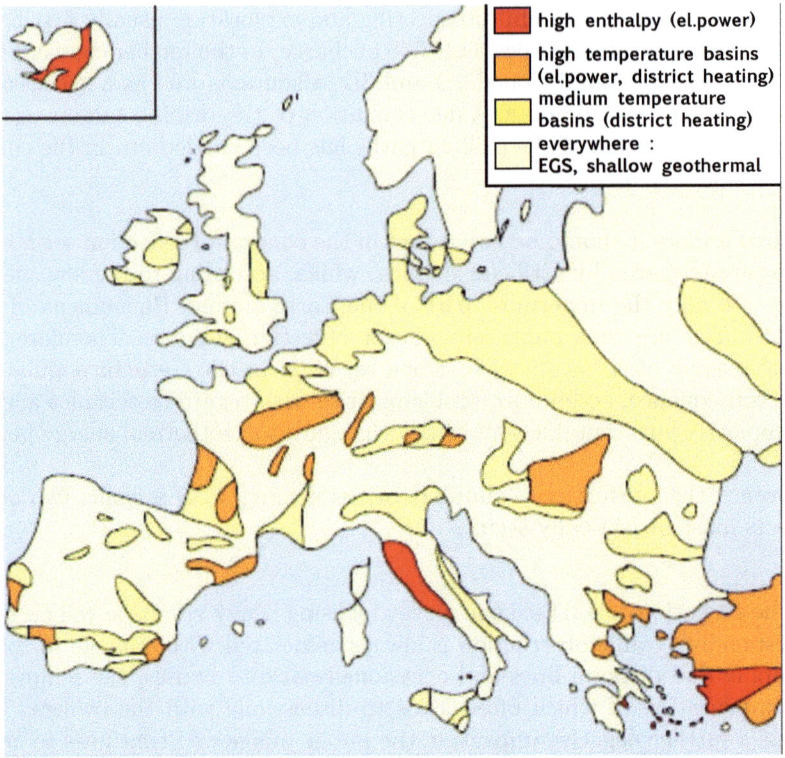

Fig. 1.13 Favourable areas for deep geothermal resources in Europe [40].

1.5 Geothermal energy, sustainability, and exploration risk

Geothermal energy is a key renewable resource with many valuable opportunities (see, e. g., [126]) such as

- *Extensive global distribution:* There is a high potential for geothermal energy production in many areas (see Figure 1.13 for an overview on the potential in Europe). Nevertheless, a joint worldwide research activity in the geothermal sector is not expedient, as the geologic and geothermal conditions vary substantially between the countries. Geothermal energy primarily requires decentralized research facilities, however, complemented by a close networking in training and education activities worldwide.
- *Independence of season:* Geothermal energy production is not dependent on any weather phenomenon (such as solar and wind influences) and is inexhaustibly available everywhere and everytime in the Earth's interior.

- *Environmentally friendly character:* Deep geothermal energy projects realize an excellent carbon balance. After the installation of a geothermal power plant the net power production is totally CO_2-free. Further valuable arguments are the creation of a local value, the prevention of a foreign policy influence, and the independence of fossil energy forms, as well as the requirement of little space.
- *Contribution to the development of diversified power:* Geothermal resources provide a reliable local energy source (free of political constraints and/or war influences) that can, at least to some extend, be used to replace the energy production based on fossil fuels.

As a consequence, the International Energy Agency report (cf. [126]) points out the perspective that geothermal resources have the potential to make a significant contribution to meeting the world's energy needs while helping to reduced emissions and to the mitigation of climate change. The global geothermal potential is enormous and the special significance of geomathematics as an interdisciplinary science can play a decisive role in the scientific consortium concerned with deep geothermal research.

The GLITNIR geothermal research report (see [183]) lists some challenges which are still valid today:

- With the rapid development of the industry, there is an *immediate need for additional expertise* (from resource development to business management), support services (drilling capacity, data/information generation and availability, technology, power generation equipment, etc.), focused capital, transparency of the assurance market, and public acceptance.
- There is still *some fragmentation in the sector to appropriately fund and develop the resource.*
- *Risk reduction including geomathematical as well as geophysical research* is absolutely indispensable. Further academic education is necessary to increase the industry talent pool. Overall project risks are higher with inexperienced scientists.
- Pressure will remain *to ramp up projects at a rapid pace*, requiring commitments from multiple players and institutions including the public sector.
- Geothermal energy is the only real renewable baseload electricity option, yet does *not get enough political support.* Other incentives for geothermal exploration are needed which could help to cover at least parts of the drilling risk for geothermal reservoirs.
- Future expansion of geothermal developments depends on exploring new fields and overcoming technical calamities in known fields that are not exploited, yet. A particular issue that should currently be addressed by the geothermal community is the development of *reliable EGS procedures* that can ensure sustainable flow rates.
- The *potential risk of induced seismicity* prevents the sustainable success of the geothermal technology and leads to the problem of social accep-

tance of the population. There is a discrepancy between the willingness of the population for more intensive utilization of "green energies" and the acceptance of neighborhoods close to geothermal facilities.

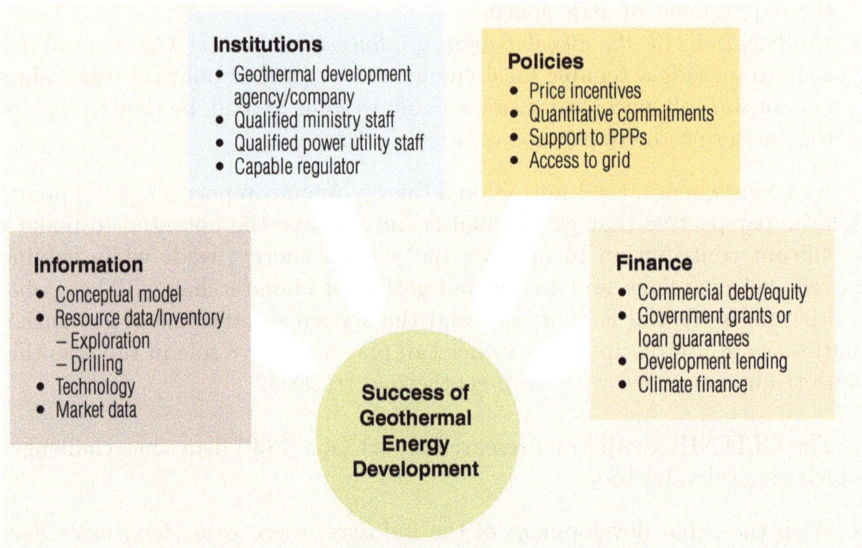

Fig. 1.14 Key elements of successful geothermal energy development due to ESMAP (cf. [84]).

We are led to the following key elements in geothermal development (see, e. g., [84]):

- Geothermal energy is well-positioned within the renewable energies.
- Advanced exploration methods are crucial to reduce risks of failure in geothermal drilling and relevant costs during the exploratory phase.
- A large and targeted geomathematical as well as geophysical research effort is continuously needed to improve the results obtained by geothermal exploration methods.
- Substantial geothermal growth could be provided by EGS technology, however, simultaneously socio-political acceptance of the population represents more and more a success factor for the realization of geothermal projects. The indicators "risk perception" and "societal and individual benefits" are decisive features.

All in all, geothermal energy becomes more and more important in the increasing demand of renewable energy production. The key elements of a successful geothermal energy development are illustrated in Figure 1.14. All developments and success factors underly a locally dependent understanding of reliable research study, and external as well as internal stakeholder management.

Geomathematics together with the scientific consortium concerned with the geothermal business will help to overcome the critical issues, i. e., minimizing risks, maximizing economic efficiency, and finally, strong need of geothermal energy retrieval, as well as socio-political acceptance. It is remarkable that, in the interplay of risk and bankability (see Figure 1.15), geomathematical components within pre-survey and exploration are both of tremendous significance and low cost.

All in all, we are led to the following conclusions: The interdependency of funding and risk found prior to a drilling project is remarkably characterized by a geomathematical/geophysical exploration phase that represents a low cost in terms of the total cost budget while maintaining a high overall significance to the project. However, despite its strong significance, until the recent past the overall contribution of the exploration phase typically only allowed for a very critical and limited meaningful assessment (see Figure 1.16).

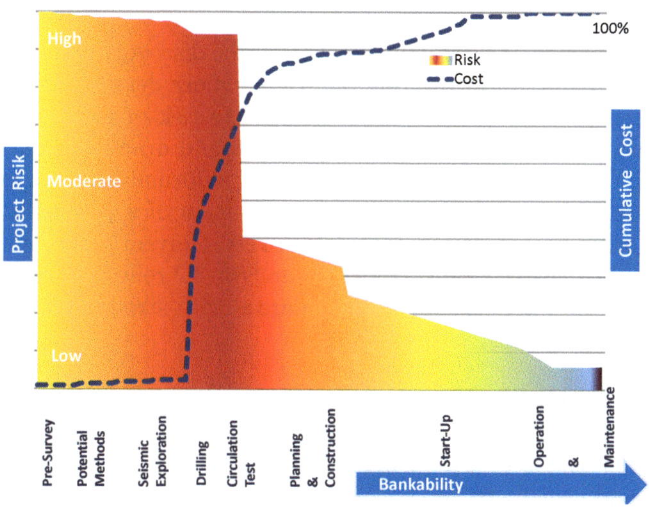

Fig. 1.15 Geothermal project risk, cumulative investment cost, and bankability following [22, 23].

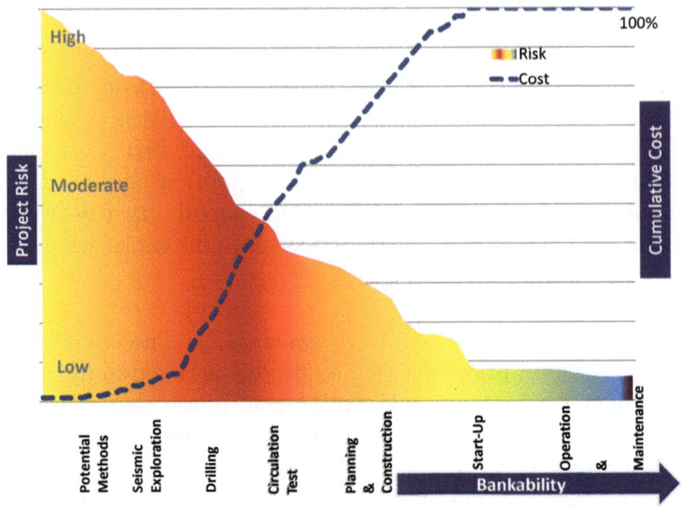

Fig. 1.16 Geothermal project risks, cumulative investment costs, and financeability (illustration from [84] as modified from [21]).

As a consequence, the reduction process of the discovery risk had to be considered as relatively uncertain (see [72, 73] for more details). In this context, it is common practice to determine the success risk of geothermal projects by incorporating various parameters (e. g., fill and drawdown data, temperature data, etc.). In addition, supplementary information from geophysical measurements must be included in the evaluation of a planned geothermal project in order to reduce the exploration risk. Currently used methods are mostly based exclusively on classical 2D and/or 3D reflection seismics (note, however, that they are not applicable in areas with strong anthropogenic influences such as mining).

In the line of Figures 1.15 and 1.16 we are immediately led to the following message for exploration bussiness: As an aid to decision-making in exploration with the aim of reducing the risk of success, people is well advised in future geothermal site determination to include less expensive geomathematical methods before carrying out costly seismic methodologies.

Chapter 2

Geomathematics as a key technology of exploration

Abstract:

Mathematics (etymologically from Greek $\mu\acute{\alpha}\theta\eta\mu\alpha =$ knowledge) intends to study topics as quantity, structure, space, and change. In this traditional understanding, geomathematics in our times is thought of being a very young science and a modern area in the realms of mathematics. However, nothing is farer from the truth (cf. [163]). Geomathematics began as man realized that he walked across a sphere-like Earth and that this observation has to be taken into account in measurements and calculations. In its character today one can only do justice to geomathematics as a key technology for geothermal exploration by briefly mentioning its significance in the superordinate geoscientific context.

2.1 Geomathematics bridging the real and virtual word

The Earth is a prime example of a complex system with interacting natural and socio-economic subsystems. Consequently, simple cause-and-effect thinking is completely unsuitable for understanding. What is required is thinking in terms of dynamic structures and an awareness of multiple, unforeseen, and certainly sometimes also unwanted effects of interventions.

The complex system Earth makes mathematics indispensable, which must be more than a collection of theories and numerical procedures. Geomathematics is nothing more than the organization of the complexities of the Earth

W. Freeden, H. Nutz, *Exploratory Potential Methods in Geothermal Power Generation*, Lecture Notes in Geosystems Mathematics and Computing,
https://doi.org/10.1007/978-3-031-54412-5_2

system. This requires descriptive thinking to clarify abstract complex issues, correct simplification of complicated interactions, an appropriate mathematical conceptual system to describe and accuracy in thinking and formulating. Geomathematics thus becomes the key science of the complex system Earth. Wherever there are data and observations, e. g., in the various scalar, vector, and tensor clusters of satellite data, or even terrestrial gravity field and/or magnetic field data, it becomes mathematical. Statistics serves for denoising, constructive approximation for methodical foundation and evaluation, special approach systems of functions give geo-relevant graphical and numerical representations, all this with mathematical algorithms.

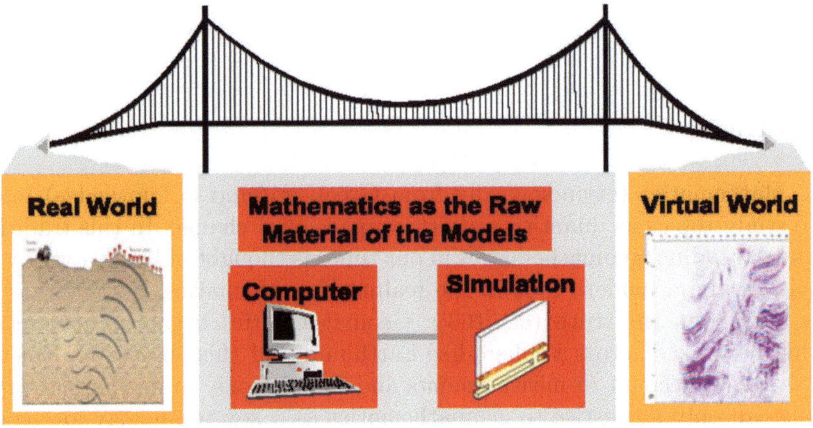

Fig. 2.1 Geomathematics as a key technology bridging the real and virtual worlds (illustration taken from [79]).

From the above it is clear that geomathematics, as the mathematics dedicated to the geosciences, is becoming increasingly important also in the field of exploration. It has the special task to span a bridge (cf. Figure 2.1) between mathematical theory and geotechnical application. The special attraction of this daughter of mathematics (cf. also [53]) is therefore the lively exchange of ideas between the group of applied mathematicians, who are more interested in model building, theoretical foundation and approximate as well as numerical problem solving, and the group of geo-engineers and -physicists, who are more familiar with measurement techniques, methodology of data analysis, implementation of routines and software application.

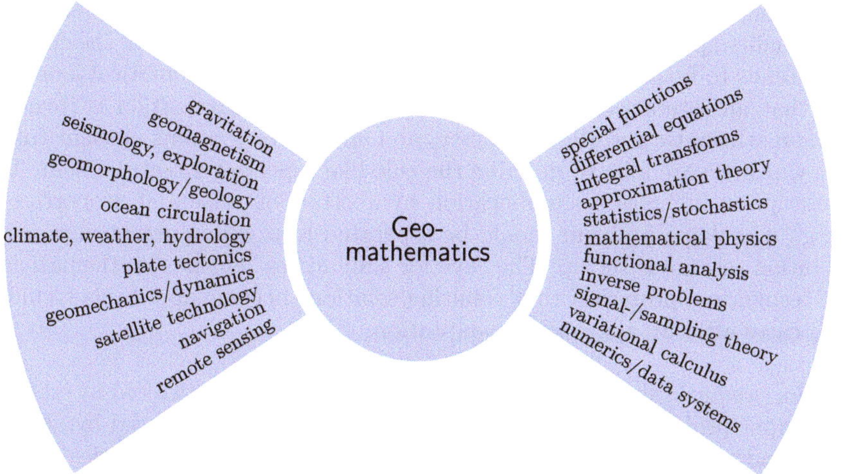

Fig. 2.2 Geomathematics, its fields of activity, and its working methods (illustration taken from [54]).

The spectrum of modern geosciences, which is the focus of geomathematics, is broad, not least because of the increasing diversity of observations. At the same time, the "box" of mathematical tools is expanding. It should be noted that geoengineering and geophysics are primarily concerned with those areas of the Earth that are inadequate or inaccessible for direct measurements. For this reason, inverse methods for evaluation become inevitable. In most cases, these methods involve measuring a physical field near the Earth's surface and then using mathematical methods to continue it into the depth ranges of interest ("downward continuation").

The basis of exploration is its measurements, i. e., scalar numbers, vectors, tensors such as distances, angles, directions, velocities, accelerations. Nowadays exploration as a measuring discipline is in great shape. Computer facilities and measurement and observation methods opened new research areas and opportunities. However, it is trademark in exploration to present measured values always together with a suitable modeling procedure for interpretation and an appropriate knowledge and estimation about reliability and accuracy. Thus, this role explains that exploration must be more than a part of geoengineering discipline concerned only with measurements. Inherently, mathematics is implied as key technology bridging the real world of measurements and the virtual world of handling datasets, modeling quantities and processes, and providing illustrations and interpretations. Once more, the result of measurements are numbers, vectors, tensors, i. e., raw material. Mathematical handling and approximation of datasets as well as modeling techniques are necessary to connect the "reality space" with the "virtuality space". In this sense, a model represents the result of the transfer, it intends to be an image of the reality, expressed in mathematical language, so that an interaction between abstraction and concretization is involved. The mathematic's world of numbers and structures contains efficient tokens by which we are able to describe the rule-like aspect of a real problem. This description includes a simplification by abstraction: essential properties of, e. g., a geologic problem should be separated from unimportant ones and a solution scheme is set up. The "eye for similarities" enables mathematicians to recognize a posteriori that resulting solutions become applicable to multiple cases after an appropriate adaptation.

So, *geomathematical exploration* is characterized by a twofold interaction. An input problem from reality space ("measurements") reduced by abstraction and transferred into virtuality space results in a mathematical output model which following a "circuit" (cf. Figure 2.3) becomes a new "concrete" input problem in reality space.

As a consequence, the ideal exploration process (circuit) for the solution process of geologic problems canonically shows the following steps to be handled (see also the approach sketched in [53]):

- *Transfer from reality to virtuality space:* Measurements and observational events in reality space lead to mathematical tokens and quantities as "raw material" for modeling and processing in virtuality space. The observational input is translated into the language of the virtuality space, i. e., mathematics, requiring close cooperation between application-oriented and mathematical scientists.
- *Retransfer from virtuality to reality space:* Appropriate analytical, algebraic, statistic, stochastic, and/or approximative as well as numerical methods must be taken into account; if necessary, new solution techniques must be proposed. The mathematical model is validated, the aim is a good accordance of model and measurement. If necessary, on the basis of new measurements, the model must be improved by use of modified "raw material".

Obviously, the benefit of a circuit is a better, faster, cheaper, and more secure problem solution on the basis of the mentioned processes of modeling, simulation, visualization, as well as reduction, decorrelation, and denoising of large amounts of data. The more measurements are available, the more one recognizes the causality between abstraction by mathematical concepts and their impact and cross-sectional importance to reality.

Evidently, the circuit in its ideal manifestation (as illustrated in Figure 2.3) has to follow an obligatory line, namely to provide an appropriate platform within which mathematically/geologically interrelated features can be

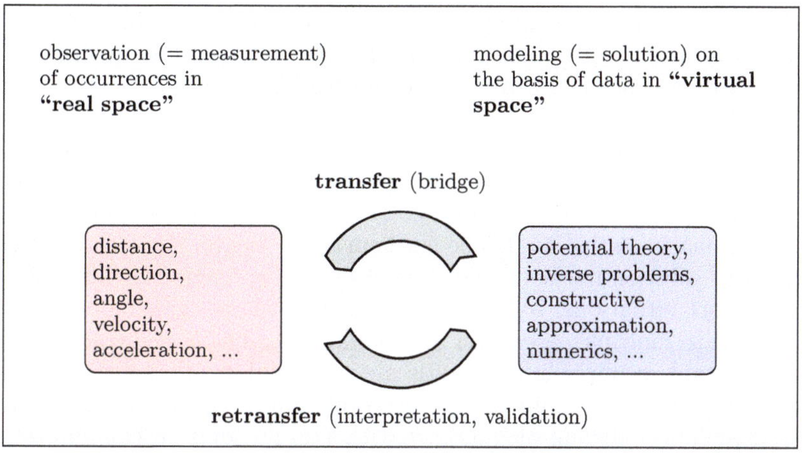

Fig. 2.3 Circuit "real space" (observation) and "virtual space" (modeling) in modified form following W. Freeden (2018), Introduction to the "Handbook of Mathematical Geodesy" (W. Freeden, M.Z. Nashed (2018), Editors).

better motivated and understood, thereby canonically leading to an interdisciplinary palette of solution procedures in exploration. In accordance with this intention, criteria must be found relative to which the scope and limitations of the various methods can be assessed. This is important both in theory and practice since there generally is no cure-all method for most of the problems in exploration.

Of course, it must be confessed that today's circuits (in the sense as depicted in Figure 2.3) turn out to be too complex in their transfer demands from reality to virtuality space and vice versa, as to be handled by only one ingenious geoscientist. Interdisciplinary solutions are urgently required as answer to an increasing complexity. In the opinion of the authors, the scientific challenge is a "geoscientific consortium" (such as GEOFÜND (PtJ FKZ 0325512A), SPE (PtJ FKZ 0324061), and SYSEXPL (PtJ FKZ 03EE4002A)), in which mathematics should not stay for geoscientists in unloved external partnership and exploration is of high quality problem-attractiveness for applied mathematicians. Indeed, the leading role of mathematics for obligations in virtuality space must be acknowledged within today's geoscience, so that mathematicians will become more enthusiastic about working on geoscientific programs. A "geoscientific consortium" reflecting the cross-sectional demands in reality as well as virtuality space is absolutely essential for a sustainable development in the future. No doubt, as this book will show, mathematicians can and should be integrated smoothly into the geoscientific phalanx instead of restricting geoengineers exclusively to measurements and mathematicians to mere service functions. Only a "geoscientific consortium" consisting of scientists with equal standing, rights, and research position will be able to promote the significance of geologic exploration in its responsibility even for society and industry.

In fact, the special importance of mathematics as an interdisciplinary science has been acknowledged increasingly within the last years in engineering, technology, economy, and commerce. This process does not remain without effects on mathematics itself. The mathematical "eye for similarities" created innovative solution methods and structures arising from very different areas and distinct situations, so that the resulting theories and models may be applicable to multiple cases after appropriate adaptation and concretization. An important example is the multiscale methodology, that enables wavelet reconstruction and decomposition, decorrelation of signatures, denoising of measurements, reducing of large datasets, etc. not only in exploration, but also in a large diversity sciences.

As a consequence, this work aims at providing innovative mathematical instruments in virtuality space in concrete adaptation to recent demands of gravity and magnetic field reflected exploration. The book is meant as a mathematical addendum to the foundations in use within today's virtuality space (cf. Figure 2.3). So, the goal of the book also is a twofold obligation:

- to make mathematicians aware of the particular mathematical developments and calamities occurring in modern concretizations of problems in exploration,
- to make geoscientists concerned with problems of exploration conscious of new tools, means, structures, methods, and procedures for handling recent measurements and observations by mathematical abstraction.

Altogether, our material presented in this book should be understood as an essential step towards modern manifestations of "geoscientific consortia" realizing the cross-sectional demands and requirements of today's circuits in a well-balanced interdisciplinary way.

2.2 Geomathematics as interdisciplinary discipline

The canonical outcome of our considerations is that *mathematics as an interdisciplinary science* can be found in almost every area of our lives. Mathematics is closely interacting with almost every other science, even medicine and parts of the arts (*"mathematization of sciences"*). The use of computers allows for the handling of complicated models for real datasets. Modeling, computation, and visualization yield reliable simulations of processes and products. Mathematics is the "raw material" for the models and the essence of each computer simulation. As the key technology, it translates the images of the real world to models of the virtual world, and vice versa (cf. Figure 2.1).

As a matter of fact, the special importance of mathematics as an interdisciplinary science has been acknowledged increasingly within the last few years in technology, economy, commerce, and in particular in exploratory technology. Certainly, as already pointed out, this process does not remain without effects on mathematics itself. New mathematical disciplines, such as scientific computing, financial and business mathematics, industrial mathematics, biomathematics, and also geomathematics have complemented the traditional disciplines. Interdisciplinarity also implies the interdisciplinary character of mathematics at school. Relations and references to other disciplines (especially informatics, physics, chemistry, biology, and also economy and geography) become more important, more interesting, and more expandable. Problem areas of mathematics become explicit and observable, and they can be visualized. Of course, this undoubtedly also holds for the Earth system.

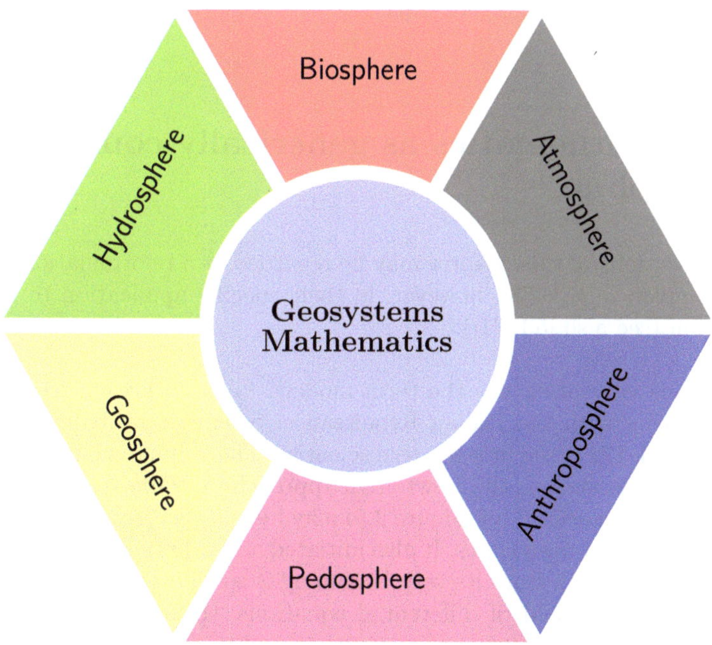

Fig. 2.4 Geosystems Mathematics as key technology penetrating the complex system Earth.

Clearly, geomathematics is closely interconnected with geostatistics and geoinformatics:

Statistics (usually seen as a basic subdiscipline of mathematics) is generally devoted to the analysis and interpretation of uncertainties caused by limited sampling of a property under study. In consequence, the focus of *geostatistics* is the development and statistical validation of models to describe the distribution in time and space of Earth sciences phenomena. Geostatistical realizations aim at integrating physical constraints, combining heterogeneous data sources, and characterizing uncertainty. Applications include a large palette of areas, for example, groundwater hydrology, air quality, land use change using terrestrial as well as satellite data. Because of both the statistical distribution of sample data and the spatial correlation among the sample data, a large variety of Earth science problems are effectively addressed using statistical procedures.

Geomathematics basically differs from *geoinformatics*. Geomathematics also deals with the further development of the language itself. Geoinformatics, however, concentrates on the design and architecture of processors and computers, data bases and programming languages, etc, in a georeflecting environment. In geomathematics, computers do not represent the objects to be studied, but instead represent technical auxiliaries for the solution of mathematical problems of georeality.

2.3 Geomathematics as generically constituted discipline

The next sections of this chapter may be regarded as a reformulation of some thoughts given in [73, 78], however, in their specific application to geologic exploration (see also [53, 71]).

Historians very often use the term modern (also, in geosciences) for the period starting with the Italian Renaissauce. Scientists probably call Isaac Newton (1643–1727) the first modern scientist. The emphasis on a systematic study of the laws such as the Newtonian approach in *"Philosophiae Naturalis Principia Mathematica"* (cf. Figure 2.5) may be regarded as the most obvious characteristic of modern times. It also initiated a new branch of mathematics, called *analysis*, which from its very beginning is so closely related to physics, for example in the field of differential equations, that some people finds it difficult to say, where mathematics starts and physics ends.

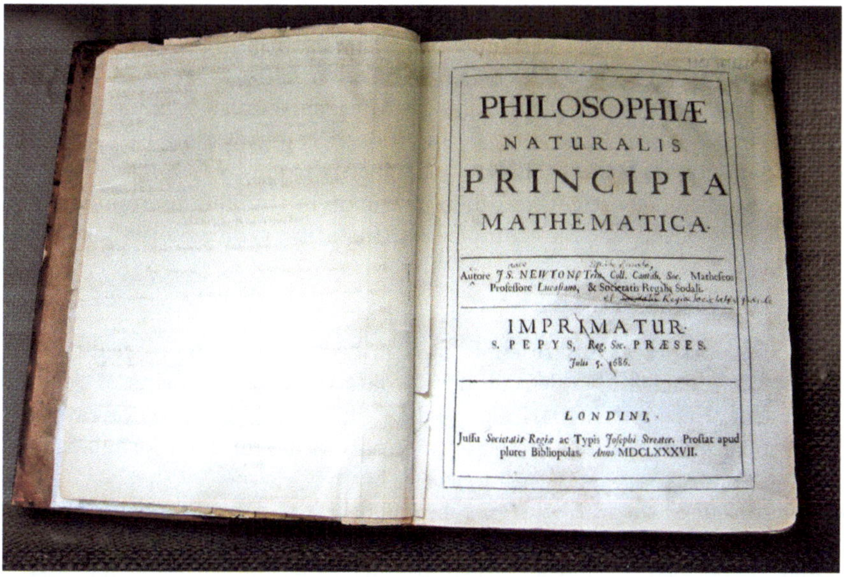

Fig. 2.5 Sir Isaac Newton's own first edition copy of his Philosophiae Naturalis Principia Mathematica with his handwritten corrections for the twentieth edition. The first edition was published under the imprint of Samuel Pepys who was president of the Royal Society. By the time of the second edition, Newton himself had become president of the Royal Society, as noted in his corrections. It has been digitised by Cambridge University Library and can be seen in the Cambridge Digital Library along with other original works by Isaac Newton. The book can be seen in the Wren Library of Trinity College, Cambridge.

The history of the interchange of ideas between the areas of mathematics and physics is a fascinating period in the history of the scientific thought. Unfortunately, it is so intricate that its complete account is still to be investigated. Nevertheless, there is a large variety of basically different ideas and concepts and an increasing change of the aspects involved in the problems. All in all, it may be concluded that one great contribution of mathematics to modern scientific progress is the setting of the pattern for the theories describing the laws of the real world ("reality"). Without mathematical theory there is no science in a modern sense.

Apart from the mathematical aspect of the value of a theory, there is a practical side to which it is even more important. Once the theory is recognized to be in accordance with an experience, it renders the experiment unneccessary. This, of course, is very important where experiments are not possible. The best example of a geodetic field which could not carry out experiments, was probably astronomical geodesy. So it was not surprising that a new approach to the features of the "real world" had its striking success when it was possible to deduce Kepler's empirical law on the movements of planets from Newton's general law of mechanics. Even more astonishing, the same rules also apply to today's space research, and nowadays satellite technology is a field in which experiments are the usual practice.

One of the best examples to show the difference in the emphasis between physics, mathematics, and engineering indeed is gravitational theory which played a major part in the history of sciences from every angle. In fact, it is well-suited for investigating different trends in thinking as represented in various fields of science, for example, electrostatics, geomagnetics, elastostatics, etc.

- The first or practically oriented step in every approach to "reality" is to gain experience and collect empirical data on certain aspects of our environment. Here, mathematics is used as a kind of registration or catalog to get concise descriptions of the laws of nature in empirical formulas.

- The second or theoretical step subsequently is to set up a system of mathematical definitions and laws from which these formulas and more results may be derived by mathematical arguing which, in turn, are intended to be verified by measurements or experiments. It must be confessed that this stage of theoretical experimentation is most confusing to a mathematician, because it uses his language and formalism without accepting his basically critical attitude. It may, however, be very exciting and stimulating because a new mathematical concept may be found.

In order to realize the second step to potential methods in exploration, gravitational potential determination was in a fortunate position, as the Newtonian theory of gravitation had already developed a mathematical status, which could be naturally extended so as to give a complete description of the laws of this field (see also [27], [124], [142]). As a consequence, potential theory as the scientific collection of ideas, concepts, and structures involving Laplace's equation gained new aspects, and it was challenged with new problems, of which the boundary value problems probably are the best known. The maximum/minimum principle shows that extremal values of a potential are to be found on the boundary of a closed body. In addition, potential theory actually guarantees that if certain values of a potential under specific consideration are given on the boundary, the potential is determined via the boundary value problem in the interior (or in the exterior when an additional regularity condition at infinity is supposed to hold true). This assertion, of course, has been checked in many experiments under different auspices in geoengineering, but naturally it cannot be verified experimentally in the generality in which it can be stated mathematically. More concretely, at a stage, where the theory is regarded as satisfactory from the physicist's point of view, it is a system of fundamental laws, definitions and problems, of which some, under certain conditions, have been solved mathematically. The problems in their full generality, however, are given to mathematics as conjectures, in a sense, to be proved. They become the object of a study of the well-posedness (in the sense of Hadamard), i. e., existence, uniqueness, stability proofs, which therefore aim at establishing the consistency of the general physical theory (see [93]). Altogether, the aspects of potential theory have changed considerably when constituents could be described by means of Laplace's equation. It could thus be observed how new physical applications developed new aspects of potential theory and the theory of partial differential equations, primarily originated on gravitational and/or magnetic developments.

In the approach of interest here, i. e., geologic exploration via a geopotential methodology, as in other scientific fields, the purely mathematical aspect is met by the engineer's need for sufficiently handling a great number of applications, which are not contained in the typical cases by which the geophysicist verified this theory. It is, for instance, certainly not enough for geoengineering to be able to calculate the "geopotential" for spheres and ellipsoids, but more complicated bodies (e. g., certain cuboids of the Earth's interior) must be taken into account. The boundary value problem, which provides a mathematical apparatus of the question of determining the geopotential from boundary values, therefore, is not only a purely mathematical subtlety, but also a problem of great practical importance. It is therefore remarkable that the link between mathematics and engineering seems to be strong, sometimes stronger than it is between mathematics and physics.

Finally it should be noted that there are two main trends in mathematics, which cannot exist one without the other: The first is to discover the laws for relations between mathematical concepts and the second is to make these structures by a process of generalization and extension as applicable and powerful as possible. As a consequence, geomathematics as a generic discipline is interpenetrated by an alternative, that may be characterized as follows:

- On the one hand, the range of mathematics contains many results which are noted for their generality like functional analysis and its manifestation in ill-posed problems.
- On the other hand, many of the famous unsolved problems have a very special nature. All these problems deal with questions of great individuality in the sense that the theories generally developed so far do not yet contain the information which is necessary to elucidate the specific ingredients of the conjectures in question.

2.4 Geomathematics and ill-posed problems

Engineering (especially, methods in geologic exploration) encouraged mathematics to deal not only with well-posed problems, but also with ill-posed problems (IPPs). As a consequence in potential theory, relevant examples, namely the "inverse gravimetry problem" as well as the "inverse magnetometry problem" became attackable, which aim at the determination of a contrast function from volume potential values on and outside the integration area. We have to note that the determination of a contrast function from internal volume potential values (cf. [62]) may be handled in well-posed framework of partial differential equations, viz. Poisson differential equation in case of gravitation, at least if the density contrast function is assumed to be sufficiently smooth and the calculation can be based on internal datasets. Geologic exploration by a gravitational/magnetic potential method, however, has to determine the interior potential and subsequently the corresponding contrast function mostly from terrestrial data (sometimes in combination with airborne and spaceborne data), i. e., the input is available only of a certain part of the boundary or outside. So, the following questions arise naturally:

- Are the specific terms of a mathematical reference space, such as distance, angle, and convergence in an infinite-dimensional Hilbert space, suitable and sufficient structures to describe solution strategies of an inverse problem?
- Is it conceivable that inverse problems like gravimetry or gradiometry from their infinite-dimensional Hilbert space formulation lead to the existence dilemma, to be considered as unsolvable?
- Is a solution - as far as available under certain restrictions - at all unique?
- Is an approximation process to the solution stable, i. e., does the inversion process lead to a bounded inverse?

As a matter of fact, in particular, the non-availability of internal data causes serious difficulties, and all properties (existence, uniqueness, stability) of well-posedness turn out to be violated in case of terrestrial and/or combined input data from the Earth's surface and its exterior space. In particular, the non-uniqueness inside is a dramatic obstacle. So, inverse gravimetry/magnetometry turn out to be severely ill-posed and cannot be handled by classical framework of potential theory as for example presented by [96, 107]. Other, more recent potential theoretic techniques have to come into play, e. g., by [13, 27, 34, 54, 61, 62, 75, 124, 142].

Looking at the developments of the interrelations between mathematics, engineering, and physics during the last century we are thus led to the conclusion: The aspects of the mathematically reflected problems of physics as well as engineering brought about that in these fields more proofs are constructive, aiming at the actual calculation at least in principle than is the case in many other fields of mathematics. However, the aspect of a mathematical theory, the way of looking at its problems, is not only changed by contacting outside areas. A theory may get innovative incentives from a change of emphasis and the discovery of a new framework based on mathematical equipment exclusively. For example, the theory of linear spaces and operators which originated from the solution of boundary value problems by Fredholm integrals (see, e. g., [57] and the references therein) is now regarded as the common heading for a large class of problems in analysis. In fact, they created a new mathematical discipline, today called *functional analysis*. This daughter of analysis offers immense progress to potential problems in exploration, since it helps "to invert the framework known from classical boundary value problems" by considering their solution expressions in form of Fredholm integral equations.

In this respect, C. Müller [127] states already in the year 1969:

> *As a matter of fact, there can be no doubt that the historic core of the discipline "functional analysis" is the Fredholm theory of integral equations and the Hilbert theory of eigenfunctions. These approaches provided the basis, that was then simplified and extended by more abstract concepts.*

Nowadays, operator theory of functional analysis in the manifestation of regularization methods (see, e. g., [62, 65]) even helps to handle *ill-posed inverse problems (IPPs)* for a large variety of applications, particularly for potential methods of geologic exploration.

In order to deal with inverse problems in exploration it is first advisible to turn to the abstract formulation of basic operator equations to get an impression of the methodological status-quo up till now. In mathematical abstraction one describes engineering or scientific problems, as inverse gravimetry or magnetometry, by an operator equation of the form

$$Ax = y, \quad x \in X, y \in Y. \tag{2.1}$$

Here we denote by x the cause from the "set" of causes X and by y the effect from the "set" of effects Y. A is an (integral) operator which connects cause and effect based on the underlying (technical or scientific) problem. As pointed out above, in exploratory obligations, X and Y are usually Hilbert spaces, i. e., complete spaces provided with distance, angle, and convergence settings.

In a direct problem one derives the effect $y \in Y$ from the measurable and thus known cause $x \in X$, that is, one computes it (directly) with the help of the underlying operator equation. If the cause x is not directly measurable or only in a very insufficient form, then one must infer from the observed, measurable effect y to its cause x. Such problems are called *inverse problems*. Inverse problems are more difficult to solve compared to direct problems because they are usually "ill-posed". It is important to recognize that this property is inherent in the problem itself and does not arise due to poor or faulty mathematical modeling. In this respect, this difficulty cannot be overcome by improved modeling of the underlying technical or scientific problem, but one needs complementary additional strategies and techniques to guarantee and improve the solvability of inverse problems.

Even simple arguments exemplarily demonstrate the (limited) performance of the basic equation (2.1) by the occurrence of so-called "ghosts" (see also [73]):

(1) Let x denote the mass density of a body, y (in the sense of Newton's theory) the measured value (output) of the resulting gravitational potential. It is possible to redistribute masses within a body in such a way that the total mass and the gravitational potential do not change. This can be done in infinitely many ways. The total mass can only be detected meaningfully if one knows additional properties of the body via internal gravitational potential values. Indeed, since Stokes in the 19th century numerous mathematicians and physicists have pointed out that the unique determination of the mass density x within an object from measured values at the surface is not possible.

(2) Another example, which also occurs in magnetic applications (see [34]), are measurements of currents y, which can be measured very accurately as output on the surface of a body. As in the case of the gravitational field of the Earth, an infinite number of different charge distributions x, which produce the same measured values y, can be specified within the body.

We denote by x the real solution ("real world solution") of the equation (2.1). Furthermore let z be a solution of the homogeneous equation $Az = 0$, which therefore produces no effect and thus has also no effect in the measurement of y. Then, because of the linearity of the operator A, we have

$$A(x+z) = \underbrace{Ax}_{=y} + \underbrace{Az}_{=0} = y. \tag{2.2}$$

In this equation, z is a so-called "ghost" when y is measured, that is, a phantom or artifact. Such ghosts occur in measuring (geo)physics and medicine, especially in the presence of only finitely many measured values, not only in the examples mentioned. The problem "ghost" leads us to the conclusion that in order to determine the real solution x one has to rely on additional information and that the operator equation (2.1) alone is not sufficient.

The occurrence of ghosts, i. e., mathematically the non-uniqueness of the solution of (2.1), is a serious calamity, but only one of the difficulties in inverse problems. We now want to address the question of what other causes lead to the fact that inverse problems are often called "ill-posed". J. Hadamard [93, 92] introduced the notion of the so-called well-posed problems. A well-posed problem shows the following three properties:

- The equation $Ax = y$ has a solution $x \in X$ for every $y \in Y$ (*existence*).
- This solution is uniquely determined (*uniqueness*).

- The inverse mapping $A^{-1} : Y \to X$ is continuous (*stability*).

In case of violation of one of these three conditions the problem is called *ill-posed* ("Ill-Posed Problem", IPP). Besides the occurrence of ghosts (i. e., uniqueness), the third condition proves to be particularly critical in practice. This is due to the fact that in measurements the data are not available exactly, but are subject to a measurement noise. If the inverse mapping is not continuous, the solution x does not depend continuously on the data y. As a consequence, smallest measurement errors can lead to drastic errors in the solution or - in the worst case - make the solution completely useless. It should be pointed out again that in the case of ill-posed inverse problems we do not obtain bad results in the solution due to an incomplete or clumsily formulated model but that this problem is inherent in the model itself. As a consequence, inverse problems are always a serious challenge due to their ill-posedness. For example, the geomathematical task of extracting geologic information of the Earth's interior from, e. g., gravimetric or gradiometric data at the Earth's surface and/or the outer space, leads independently of the targeted depth position to an ill-posed inverse problem in the sense defined above, where even all three conditions (i. e. existence, uniqueness and stability) are violated.

Now, how can we still obtain usable information despite the ill-posedness of a problem (2.1)? Indeed, for the treatment of ill-posed inverse problems, a large number of methods have been developed in the last decades. The basic idea in most of the solution methods (for approximate solution) of the IPP is to construct a "solution" which is physically acceptable as a reasonable approximation and numerically stable enough. The basic problem in modeling ill-posed problems is the following relationship: The more accurately the mathematical model describes the IPP, the worse the "condition number" of the associated computational problem, i. e., the more susceptible the procedure is to errors. The difficulty with ill-posed problems is, in particular, to integrate additional information, compromises or new perspectives helpfully and purposefully into the existing solution in such a way that the accuracy of the solution is actually improved. Usually, as already mentioned, one talks about of the so-called "regularization of an ill-posed problem" for the description of the various approaches to circumvent this problem of missing continuous dependence (as well as the study of existence and uniqueness statements). Roughly speaking, this involves analyzing the IPP by handling corresponding associated well-posed problems, i. e., a family (usually a sequence or system) of well-posed problems that provide meaningful answers to the ill-posed problem. Here, three aspects are of relevance in regularization (cf. [62, 63, 64, 65]):

1. solution and reconstruction strategy,
2. regularization-approximation scheme,

3. algorithms for regularization.

We distinguish different methods and results:

Ad 1: The *strategy of the solution and reconstruction of ill-posed problems* contains at least one of the following intuitive ideas (for more details we refer to the present textbooks of inverse problem theory):

- Adjustment of the default to the accuracy of a solution u, e. g.
  - ε-accuracy of the solution: $\|Au-y\| \leq \varepsilon$ for given ε,
  - Quasi-solution: $\|Au-y\| \leq \|Ax-y\|$ for all $x \in M$, where M is a given subset of the domain of definition of A,
  - Least squares solution $\|Ax-y\|_Y = \inf\{\|Az-y\|_Y : z \in X\}$,
  - Best-approximating solution $x^\dagger = A^\dagger y$ with $\|x^\dagger\|_X = \inf\{\|z\|_X : z$ is least-squares solution of $Ax = y\}$,
  - aso.
- Modification of the operator equation or the problem itself (for example, by constructing mollifiers for the kernels in the integral equation under discussion (see [34, 54]),
- Specification of the type of noise and its inclusion in the proposed solution.

The philosophy of solution leads to the use of algebraic solutions vs. solutions in function spaces, statistical solutions vs. deterministic solutions, strong vs. weak noise, etc.

Ad 2: With respect to *regularization-approximation schemes*, we refer to a number of methods that are used in ill-posed inverse problems, such as Tikhonov-regularization, projection methods, multiscale methods, iterative approximation, filter and mollifier methods, etc. A detailed description of these methods is omitted here.

Ad 3: By implementing these schemes, they lead to certain *algorithms for regularization*. Unfortunately, this includes the determination of an appropriate value of a certain parameter, associated with the scheme (e. g., regularization parameter, mesh size, dimension of the subspace in the projection scheme, determination of the degree of multiscale spaces, decorrelation level, classification of noise, etc.). This turns out to be a nontrivial problem, since there is consequently a trade-off between accuracy and numerical stability. Such a conflict does not usually arise in well-posed problems.

Currently, seen from the standpoint of mathematical and numerical analysis, one can roughly group "regularization methods" into several (canonically overlapping) categories (for more details see [68]):

- *Regularization methods in function spaces* is one category. This includes Tikhonov-type regularization (see, e. g., [132, 173]), the method of quasi-reversibility, the use for certain function spaces such as scale spaces in multiresolutions, the method of generalized inverses (pseudoinverses), e. g., in reproducing kernel Hilbert spaces (cf. [131, 137, 138]), and multiscale wavelet regularization (see, e. g., [52, 62, 65, 76]).

- *Resolution of ill-posed problems by "control of dimensionality"* is another category (see, e. g., [135] and the literature therein). This includes projection methods and moment-discretization schemes. The success of these methods hinges on the possibility of obtaining an approximate solution while keeping the dimensionality of the finite-dimensional problem within the "range of numerical stability". It also depends on deriving error estimates for the approximate solutions that is crucial to the control of the dimensionality.

- A third category is formed by *iterative methods* (see, e. g., [105, 106]) which can be applied either to the problem in function spaces or to a discrete version of it. The crucial ingredient in iterative methods is to stop the iteration before instability creeps into the process. Thus iterative methods have to be modified or accelerated so as to provide a desirable accuracy by the time a stopping rule is applied.

- A fourth category constitutes *filter methods*. Filter methods refer to procedures where, for example, values producing highly oscillatory solutions are eliminated. Various "lowpass" filters can, of course, be used. They are also crucial for the determination of a stopping rule. *Mollifiers* (see, e. g., [42, 43, 54, 63, 64]) are known as smooth filter functions with special properties to create sequences of smooth functions approximating a nonsmooth or singular function. Thus, we compromise by changing the problem into a more well-posed one, namely that of trying to determine a mollified version of the solution. Once more, the heuristic motivation often is that the trouble comes from high frequency components of the data and of the solution, which are damped out by mollification. Mollification is usually used when the ill-posed problem is "too ill-posed" to apply the standard regularization methods. As a matter of fact, as we shall see later, in Earth's exploration by potential methods mollifier decorrelation brings out the transitional zones of geologic strata by multiscale wavelet techniques (for more details see also [54] for inverse gravimetry and [34] for inverse magnetometry).

- The root of the *Backus-Gilbert method (BG method)* was geophysical (cf. [14, 15, 16]). The characterization involved in the model is known as moment problem in the mathemetical literature. The BG method can be thought of as resulting from discretizing an integral equation of the first kind. Where other regularization methods, such as the frequently used Tikhonov regularization method, seek to impose smoothness constraints on the solution, the BG method instead realizes stability constraints (cf. [185]). As a consequence, the solution is varying as little as possible if the input data were resampled multiple times. The common feature between the mollifier and the BG method is that an approximate inverse is determined independently from the right side of the equation.

- The usual determination of parameters by inverse analysis assumes that the parameters are deterministic, that is that they are unknown but constants. The statistical approach to estimating parameters assumes that the parameters are random and that some knowledge of their statistical properties is known via a prior probability density distribution. Originally, *statistical regularization* is the name given to a form of estimation which resembles the usual Tikhonov-Phillips regularization method in which the regularization parameter is related to the variance of the prior. In modern nomenclature, the designation *stochastic regularization* is used to reduce the sensitivity to the degree of uncertainty in the parameters implying an estimation of the data, for example, a mean square estimation of a parameter which has a Gaussian distribution when the input data is also Gaussian distributed.

- Linear models for nonlinear relationships usually describe the reality in a very limited way. Thus it is required to work with *nonlinear concepts*. Because of the diversity of nonlinear phenomena, there is no closed inverse theory. As a consequence, nonlinear inverse problems are mainly based on the specific properties of the individual case.

Each method for resolution involves a critical "parameter" whose "optimal value" is crucial to the amenability and numerical implementation of the method. In *Tikhonov-type regularization* it is the regularization parameter, or more generally the choice of the regularization operator. In *multiscale methods* it is the scale parameter to determine the scale space in which multiresolution is realizable relative to the data width and noise. In *projection* and other discrete methods, it is the optimal dimension of the approximating subspaces. *Mollifier regularization* of integral operators (cf. [34, 54]) uses a kernel function that, convolved with a particular function, yields a filtered function, which is "close" to the function but "smoother". Usually, a mollifier is not only one kernel function but a sequence, or usually a one-parameter (i. e., scale dependent) family of kernels (also known as an approximate identity), so that the kernels of a mollifier have to tend to the Dirac kernel. As a consequence, mollifier kernels become more and more space-localized, and this make them extremely attractive for numerical purposes. In addition, filtering (convolution) with mollifier wavelets constituting the difference of two subsequent mollifier kernels provides bandpass information which, with increasing scales, deliver better and better details up to a certain scale. In discretization methods it is the choice of the mesh size beyond which a further refinement will lead to instability. In *iterative methods* it is the moment at which one should terminate the iteration (i. e., it is the number of iterates). In *filter-truncation methods* it is the number of terms to be included, etc. Our work intends to delineate suitable principles that quantify the choice of the parameter, the type of estimates, and a priori information that are needed to arrive at an "optimal" value for this parameter.

There are *more additional characteristics for ill-posed operator equations* $Ax = y$, where X and Y metric spaces and $A : X \to Y$ is a continuous injection: For a long time, the use of *a-priori-boundaries* (or more generally a-priori-information) in solving an ill-posed problem has been considered essential to improve continuous dependency (i. e., to achieve a regularization effect). At the beginning of the studies of ill-posed problems, Tikhonov made the crucial observation (see [171, 172]) that the restriction to a compact set ensures well-posedness. More precisely, if X and Y are metric spaces, $A : X \to Y$ is a continuous injection and $C \subset X$ is compact, then $A^{-1} : F(C) \to C$ is continuous.

The method of successive approximation gradients and relative iterative methods can be used to search for an approximate solution of ill-posed problems. A convergence analysis of some iterative methods for integral and operator equations of the first kind and for equations of the second kind at a characteristic value can be found, for example, in [106]. These methods include as a special case what is sometimes called Landweber-iteration (see [167]), although this is in fact identical to the method of successive approximation applied to normal equations, i. e., we have $x_{n+1} = x_n - A^* A x_n - A^* g$ for the operator equation $Ax = y$, where A^* is the adjoint operator of A. When an iterative method is applied to numerically approximate a problem (whether well- or ill-posed), a stopping rule must be developed to terminate the method. Rounding errors and the tolerance of the approximation error (i. e., what is considered "zero" from the point of view of the approximation) must be taken into account. For ill-posed problems the termination rule is clearly more complicated due to the numerical instability caused by the ill-posedness. For the case where y is noisy, i. e., $y^\varepsilon = y + u$ with $\|u\| \le \varepsilon$, the approximations

$$x_{n+1} = x_n - \beta(A^* A x_n - A^* y - A^* u) \tag{2.3}$$

(where β is a parameter for which this iteration converges when $u = 0$) need not converge to the generalized inverse (or pseudoinverse) $A^\dagger$. The same is true for the method of steepest descent when γ is replaced by γ_n, which is the value at which $\|Ax_{n+1} - y^\varepsilon\|$ is minimized. Naturally, the termination criterion depends on the residual as an observable quantity. If one chooses $\delta > 0$ (termination level), one obtains at the beginning of the iteration

$$\|Ax_n - y^\varepsilon\|_Y > \delta, n < m, \|Ax_m - y^\varepsilon\|_Y \le \delta. \tag{2.4}$$

We assume that the stop level $\delta = \delta(\varepsilon)$ of this iterative process for $\delta(\varepsilon) \to 0$ satisfies the condition $\delta(\varepsilon) > \varepsilon$. Then it is easy to show that

$$\lim_{\varepsilon \to 0} \sup_{\|u\| \leq \varepsilon} \|x_M - A^\dagger y\| = 0, \tag{2.5}$$

where $x_M = x_M(\delta, u, x_0)$ is the approximate solution obtained from (2.3) with stop level δ. Thus, it is shown that one can apply the iterative process as a regularization scheme under these conditions. The result (2.5) means that the family $\{B_n\}$, where B_n is the operator (i. e., the iteration rule) that maps y to x_{n+1}, is a regularizer. In this case $\frac{1}{n}$ takes the role of the parameter γ.

Furthermore, let us mention that in iterative methods, the function $x_n(\delta(\varepsilon), u, x_0)$ can be used as an "optimal" approximation solution that represents a tradeoff between accuracy and stability. Since x_n is a linear combination of a finite number of functions (introduced recursively in [116]), it follows that regularizing iterative schemes can be considered in the context of resolution by controlling dimensionality. In this case the functions, however, are not given a priori, but they are recursively generated by repeated filtering of an initial approximation x_0. The focus is then on the filter operators by using the adjoint and on the choice of x_0.

One approach to solving ill-posed problems is to seek an approximation of the solution (of the inverse problem or of the identification problem) by a linear combination of a "small" number of prudently chosen functions. The problem then reduces to the best approximation with "few" parameters. For certain ill-posed problems a clever choice of functions can be made.

The idea of *regularization by discretization or projection methods* is very popular in numerical analysis of ill-posed inverse problems. The idea is simpler (in principle) than the sophisticated method of Tikhonov-regularization. The application of such methods predates the introduction of regularization. On this occasion, it should not go unmentioned that Maxwell was among the first scientists to formulate and solve an inverse problem in 1879 while trying to calculate the capacity of a square metal plate using integral equations. His numerical approach was basically collocation. The construction of antenna systems and numerous electromagnetic apparatuses were also linked to the development of projection methods.

A framework for the reduction of inverse problems and identification problems to finite-dimensional problems exists within the classical projection methods. We choose a family of finite dimensional subspaces $X_n \subset X$, $Y_n' \subset Y'$ which are associated to the problem $Ax = y$, where $A : X \to Y$ and Y' is the dual space of Y (i. e., the space of all continuous linear mappings from Y to $\mathbb{R}$). We assume for simplicity that the range $\mathcal{R}(A)$ is dense in Y and $y \in \mathcal{R}(A)$. The approximations $x_n \in X_n$ for $x = A^{-1}y$ are given by the condition

$$\psi_n(Ax_n - y) = 0, \quad \psi_n \in Y_n'. \tag{2.6}$$

Using noisy data, i. e., substituting y for y^ε and using the projection method we have the following basic estimation

$$\|x_n^\varepsilon - x\|_X \leq (1 + \|P_n\|_X) d(x, X_n) + \|Q_n\|_X \varepsilon. \tag{2.7}$$

Here the linear mappings $P_n : X \to X_n$ and $Q_n : Y \to X_n$ are defined by $\psi_n A P_n = \psi_n A$ and $\psi_n A Q_n = \psi_n$ for all $\psi_n \in Y_n'$ and with $d(x, X_n) = \inf\{\|x - u\|_X : u \in X_n\}$. The first part of the right side of (2.7) is the *projection (or discretization) error*. If $\|P_n\|$ is uniformly bounded in n, i. e., if the method is quasi-optimal in the language of finite-element projection, then the order is independent of ψ_n and depends only on how well x can be approximated in X_n. The second part of (2.7) is the *error due to noise* amplified by the ill-posedness. One can show that $\|Q_n\| \geq a(n)$, where

$$a(n) = \sup\{\|u\| : u \in X_n, \|Au\| = 1\} \tag{2.8}$$

depends on X_n but not on ψ_n. For ill-posed problems, $a(n) \to \infty$ if $n \to \infty$, because the norm of the inverse matrix arising from the system of algebraic linear equations resulting from (2.7) is not uniformly bounded. If $\|Q_n\|$ with $n \to \infty$ does not grow faster than $a(n)$, then the projection scheme is called robust. If a projection scheme is both quasi-optimal and robust, then

$$\|x_n^\varepsilon - x\|_X \leq C\{d(x, X_n) + \varepsilon a(n)\}. \tag{2.9}$$

This estimate cannot be improved by a more clever choice of ψ_n. Furthermore, a compromise can be achieved by minimizing the right-hand side with respect to n. Numerous concrete realizations within this scheme of projection methods are obtained, for example, in [141]. In [108] a detailed analysis of different projection methods is made.

Projection methods (using spline, wavelet or finite element functions) can be applied, as in the case described above, either directly to the IPP or to the regularized problem, i. e., to the problem of the minimization of the functional $J(x) = \|Ax - y\|^2 + \gamma\|LX\|^2$, $a > 0$. Various convergence results are obtained at the early stage of the IPP theory in [119, 141, 164]. In the bibliography there are also further useful references to projection and finite element methods for IPPs.

The momentum discretization- (or semi-discretization-) method for integral equations of the first kind and for IPPs in Hilbert spaces with reproducing kernel structure (see [131, 132, 137, 138] is in the sense of a projection method both quasi-optimal as well as robust. It is particularly suitable if only at finitely many points values of the data function are known. These properties, in conjunction with convergence and commutativity properties shown in [133] are responsible for the uncomplicated behavior of these methods in numerical implementation.

There is now a large number of textbooks on IPPs with comprehensive and in-depth results, e. g. [5, 18, 42, 43, 44, 89, 108, 117, 128, 130, 136], to name a few.

All in all, we are led to the following conclusion: Inverse problems, due to their mathematical specificity, never provide absolute certainty in their solution. The reason lies in the specific character of inverse problems (such as gravimetric and/or gradiometric inverse problems), which have to draw the conclusion from indirectly measured effect to the real cause for an area of the Earth, which is completely beyond direct observation when entering a project. This circumstance inevitably implies the "occurrence of methodically conditioned ghosts" in the solution of an inverse problem, which represent "essential horrors" of any solution effort. The ideal state must be to make these methodically conditioned ghosts of an inverse method complementary visible by an alternative method. Only in complete complementarity the residual risk may be reduced to the extent that false statements and misinterpretations could be avoided. Even more, for the inverse data evaluation it is to be stated that the traditional mathematical procedures will be able in the future neither in theory nor in numerics to master the new datasets - in particular under the important aspect of a deepened spatial-temporal treatment. Rather, what is required is a processing of the geoscientifically relevant quantities into constituent building blocks that is characterized by three essential features: good approximation properties, suitable decorrelation capability, and fast as well as stable algorithms. These features are the key to a variety of skills, in particular data compression and transmission, denoising, and selective multiresolution of, for example, geologic structures.

It remains to be said: Geomathematics for the treatment of inverse problems is and remains a key technology also in the treatment of inverse problems of geologic exploration. Even if great progress has been made in the last decades and can be expected for the future, the mathematical cycle (cf. [53]) consisting of

- *mathematical model building,*
- *mathematical analysis of the constituents of "well-posedness",*
- *development and implementation of a mathematical solution procedure,*
- *back transfer into application*

will remain indispensable for the foreseeable future.

Chapter 3

Potential data and basic mathematical modeling

Abstract:

Potential exploration methods aim at locating geothermal resources and water reservoirs, minerals, faults, and/or petroleum resources. Potential field surveys for detecting geothermal resources are relatively cheap, non-invasive, and non-destructive environmentally speaking. They are passive, i. e., no energy needs to be put into the ground in order to acquire exploration data. The portable instruments (gravimeter and magnetometer) also permit walking traverses. So, the goal of studying potential fields is to provide a better understanding of the subsurface geology under low(er) expenses. However, gravity and magnetic prospecting are affected by the fact that the measured signal is a composite of the contributions from all depths. In fact, the uniqueness problem in gravimetry and magnetometry is the most serious obstacle, when only terrestrial data are obtainable. As a consequence, superpositions over all depths can only be separated suitably, if additional information is available, e. g., from earlier geologic campaigns or borehole activities. Moreover, innovative decorrelation methodologies of the geologic layers are inevitable.

3.1 Exploratory requirements and potential data

The omnipresent gravitational and magnetic forces of our planet Earth present themselves to different degrees depending on the location. These minimal fluctuations (more precisely, anomalies from a "normal model", that will not be specified here in more detail) are highly interesting for researchers in

W. Freeden, H. Nutz, *Exploratory Potential Methods in Geothermal Power Generation*, Lecture Notes in Geosystems Mathematics and Computing,
https://doi.org/10.1007/978-3-031-54412-5_3

a variety of scientific disciplines, particularly in geologic exploration. Exploration means the geologic identification of cost-effective drilling regions. The decisive factor is the discovery of geologic layers, especially fault patterns, and the accompanying disturbances and discontinuities in the interesting area of the underground.

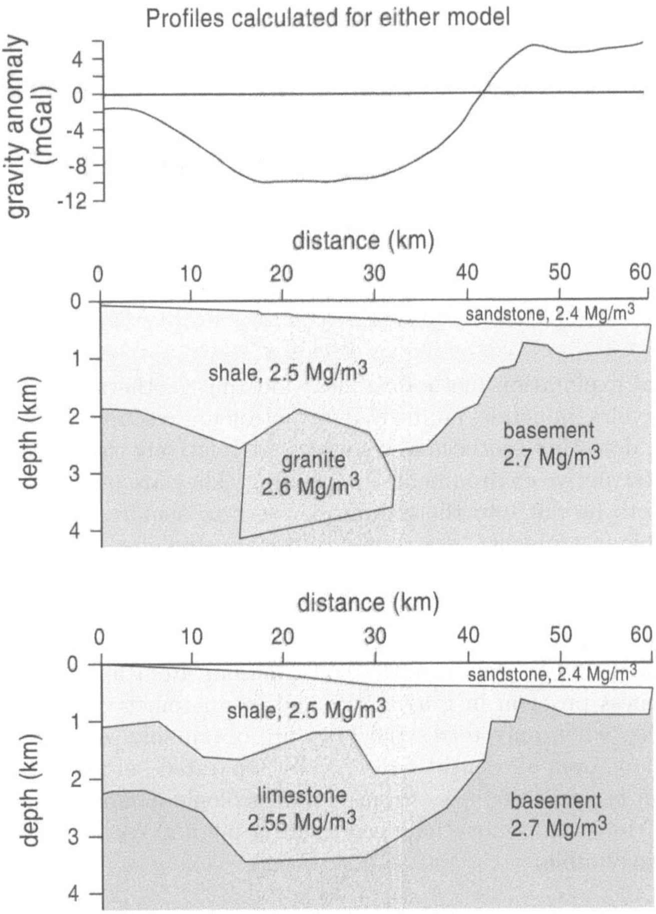

Fig. 3.1 Illustration of the mass profile of different density distributions in the subsurface and the resulting gravitational effect (modified figure according to [177]). Analogous uniqueness problems arise when determining the distribution of magnetization from terrestrial magnetic field functionals (see, e. g., [153]).

The rationale in most methods for resolution (approximate solvability) of ill-posed inverse problems of determining the contrast function F from potential input data V (such as inverse gravimetry or inverse magnetometry) is to construct a "solution" that is acceptable physically as a meaningful approximation and is sufficiently stable from the computational standpoint, hence, an emphasis is put on the distinction between "solution" and "resolution". The main dilemma of modeling ill-posed problems (IPPs) is that the closer the mathematical model describes the IPP, the more sensitive is the associated computational problem. As a consequence, all components of potential data characteristics, i. e., quantity, quality, location, and density distribution play an essential role.

So, even at this early stage (cf. [54]), two major difficulties arising from potential data should be mentioned and kept in mind:

- The measurement of terrestrial gravitational intensities in no way implies that the arrangement of masses inside the Earth is unique (see Figure 3.1 for a well-known demonstration).

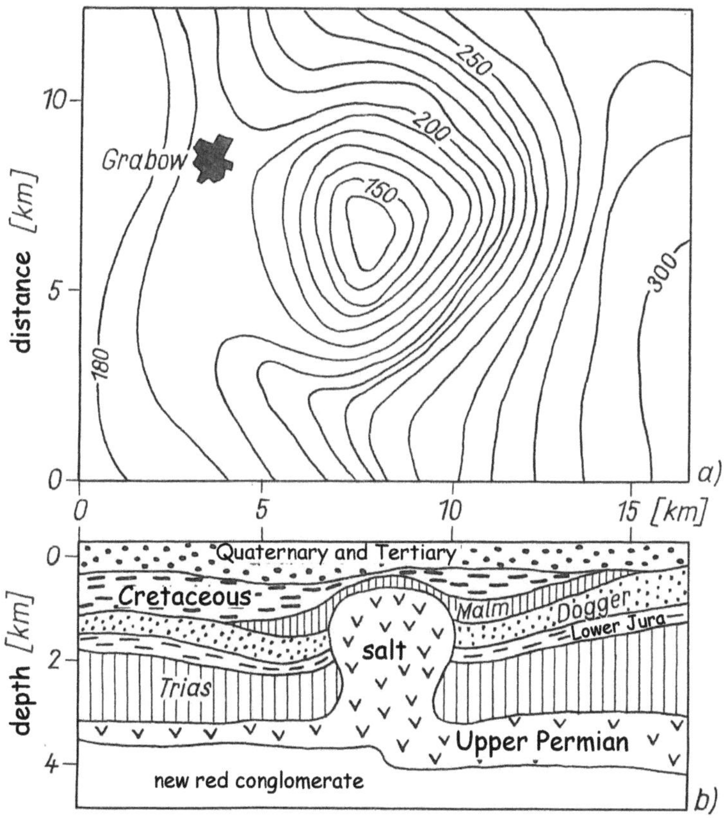

Fig. 3.2 Top: Gravity effect in $[\mu m\ s^{-1}]$ of the salt dome Werle (Mecklenburg, Germany); bottom: Geologic vertical profile (according to [104], modified illustration with kind permission of Teubner, Leipzig).

- According to practical experience in the past, for a successful application of gravimetric and/or magnetometric investigations, it is essential that the geologic structures to be determined stand out from their surroundings by clear density contrasts, hence, a common prerequisite for the successful application of today's gravimetric investigation methods is that the geologic structures to be detected must show sharp contrasts in density. In Figure 3.2 this is shown exemplarily for a salt dome by gravimetric isolines. Since the density in the salt dome is distinctly contrasted with the surrounding rock, the gravimetric signals show a characteristic isoline image of a minimum in the center of the salt dome. Such structures, but almost only such structures, can be handled and analyzed more precisely with traditional gravimetric exploration methods, so that suitable conclusions about the salt dome can be drawn from the gravimetric signals.

The geomathematical methodologies presented in this book, however, claim to be able to detect the transition surfaces between less distinct geologic formations, e. g. areas outside the salt dome. A wavelet decorrelation, i. e., a multiscale decomposition of gravitational signatures by wavelets, is a decisive tool in this respect, so that layer boundaries of geologic formations become visible and thus can be planned. Nevertheless, the wavelet type must be specified carefully, the wavelets should be close to the physical reality, for example, certain differences of "mollified monopoles or dipoles".

In order to reduce the described challenges of a 3D seismic survey and at the same time to provide a proper possibility for validation (if deemed necessary), the analysis of geopotential (anomalies of the) gravity and magnetic field is an inexpensive possibility. Without an extensive permitting process and large administrative obligations, potential data can be collected almost independently of seasons, thereby providing subsurface information that at least allows the area to be covered to be concentrated in interesting target areas or even allows a project to be terminated early without having already invested a large six-digit sum of money. At the same time, conditions are created (in advance of a seismic survey) to compare the interpretation of seismic signals with results from the analysis of the gravity and magnetic field and to reconcile discrepancies with regard to risk mitigation. However, the desired goal in future should be to renounce seismic activities completely.

It is self-evident that the success or failure of a geotermal project is essentially determined by the available exploration dataset. Here it must be confessed that geologic exploration by potential methods needs a denser local potential data distribution than offered by geodesists and geophysicists by earlier measurement campains. In the context of the projects SPE (PtJ FKZ 0324061) and SYSEXPL (PtJ FKZ 03EE4002A), this circumstance was taken into account by a series of additional own measurement activities using the facilities of the company CBM, Bexbach, Saarland, Germany.

An important aspect of the use of gravimetry and magnetometry in exploration is the consideration of the *measurement accuracy*. It must be ensured that the gravitational and magnetic field (anomalies) to be modeled are at least in the order of magnitude of the measurement accuracy. It should be noted in this context that due to the development of more and more powerful gravimeters and magnetometers also much weaker potential anomalies can be detected and thus modeled in future by means of advanced geomathematical decorrelation methods, so that potential methods will inevitably move more into the focus of exploration, e. g., in geothermal energy, due to increased measurement and observation as well as modeling scenarios.

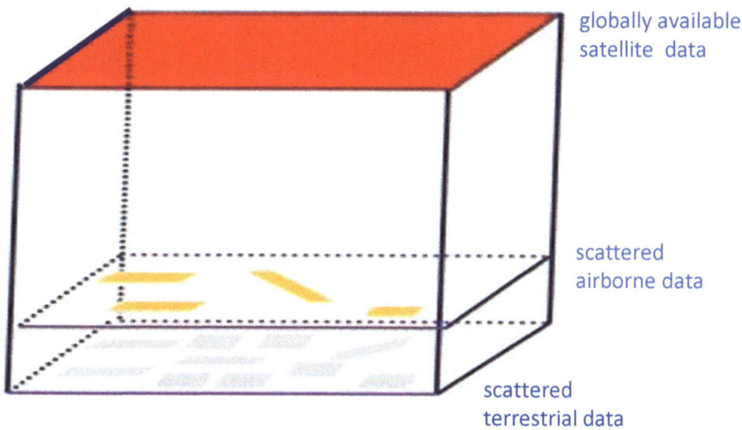

Fig. 3.3 Schematic illustration of the height-dependent "patchwork" of available data sets of the gravity and magnetic fields.

Unfortunately, it should be emphasized that in reality the existing terrestrial data situation of the Earth's gravity and magnetic field is far from homogeneous and suitably dense distribution and must be supplemented by own measurement campaigns. Moreover, the quality of existing data of the geopotentials to be considered varies significantly. At present and in the foreseeable future a dense coverage with sufficient data of corresponding accuracy is far from a satisfactory situation and mathematical substitute solutions must be sought indispensably with the help of the underlying partial differential equations as mathematically/physically motivated bridges. Even more, for data supplementation and for numerical stabilization namely at the boundaries of an investigation area, i. e., for the elimination of artificial oscillation phenomena (like "Gibbs phenomena"), airborne and/or spaceborne data are indispensable. Overall, the terrestrial data situation requires special attention with respect to homogenization and unification in order to avoid undesired undulation phenomena within the modeling process of the scattered data namely in transitional areas of the "patchwork" (see Figure 3.3).

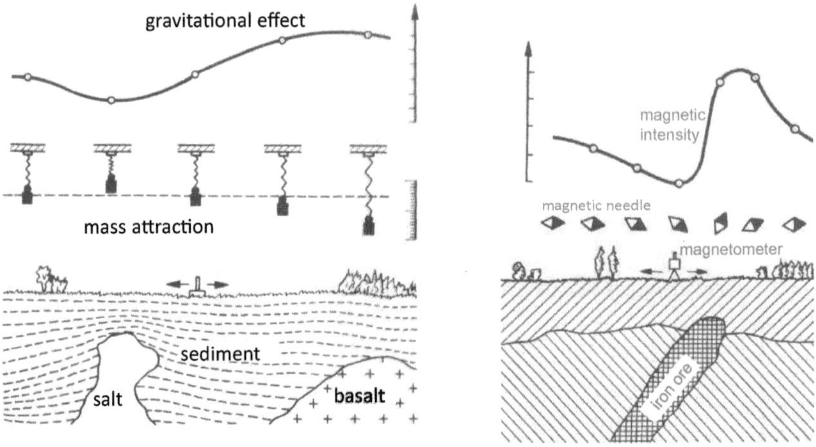

Fig. 3.4 Principle of gravimetry (left) resp. magnetometry (right) (taken from [104] with kind permission of the publisher).

Both gravity and magnetic exploration methods are non-invasive methods because they investigate geologic conditions by measuring at the surface without impacting internal structures (see Figure 3.4). Gravito-magneto-based methods are thus socially acceptable. Despite the required dense measurement grids and the measurement accuracies to be guaranteed, gravito-magneto-based methods are also more cost-effective than reflection seismics. Furthermore, areas where large-scale salt, ore or coal mining and the resulting caverns and loosening zones or dense development in urban areas make the use of classical geophysical instruments such as 3D reflection seismics impossible or seriously prove to be problematic for exploration. In fact, caverns no longer allow the acoustic and/or elastic waves to penetrate to the layers of interest. So, in heavily anthropogenically influenced areas, reflection seismics is practically inapplicable. In these cases, alternative geoscientific methods must necessarily be developed and tested for their usefulness. All in all, one is then obliged to base the development and use of new mathematical-physical methods for the analysis of a geothermal system (anthropogenically influenced by mining) on combined gravimetric and magnetic data, in which additionally available seismic information can and should be interwoven.

It is helpful and useful to develop and consider the application of gravitational and magnetic field methods in parallel, although there are of course significant differences (cf. [34, 54, 55]). This can be seen especially in the fact that the gravitational field points to the center of mass of the Earth, while the magnetic field can be understood as a bar magnet with center in the center of the Earth (cf. Figure 3.5).

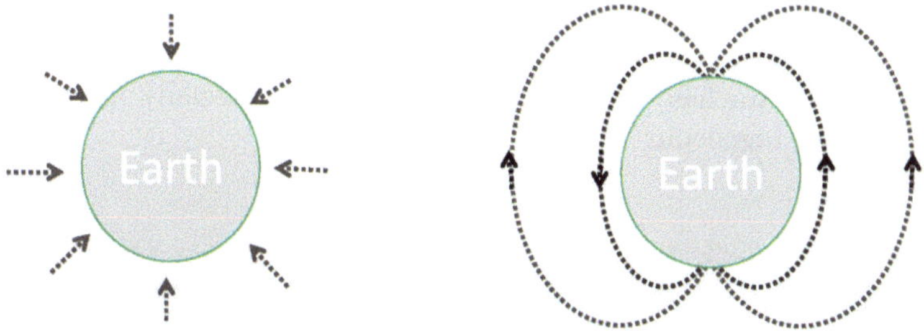

Fig. 3.5 The gravitational field lines (left) are monopolar, i. e., they point to the center of mass of the Earth, while the magnetic field lines (right) are dipolar (survey drawing).

There are, unlike gravity, no magnetic monopoles and therefore dipoles (and occasionally higher order poles such as quadrupoles, etc.) are the main constituents of magnetic fields. Magnetic poles can be repulsive or attractive, unlike in the case of gravity. Consequently, the geophysically based construction of gravimetrically and magnetometrically reflected mollifier wavelets must reflect, respectively, the monopole and dipole character of the potentials under study, and the mathematical treatment of decorrelation methods in gravimetry and magnetometry is different.

As a consequence, in exploration practice, gravity and magnetic field inversion are usually treated separately in order to obtain density and magnetization independently via different kinds of modeling processes. Nevertheless, with the aid of Poisson's equation (see, e. g., [27]), it can be shown that for any subset of a body with uniform magnetization as well as uniform density, there is a correlation. If a body has uniform magnetization and density, the magnetic potential is proportional to the gravitational field component in the direction of the magnetization (see, e. g., [27, 57]). This fact allows (at least from its physical approach) gravito-magneto-combined inversion. Thus, testwise, there is the possibility to determine deviations from uniformity and thus to obtain more reliable combined information from gravitational/magnetic anomalies (cf. [57]). After the evaluation of global gradiometer data of the gravitational and magnetic field from satellite measurements (e. g., GOCE data, SWARM data) during the initial phase of a geophysical investigation of the subsurface, a mathematically and physically consistent combination of these data with airborne (if available) and terrestrial information (if available) is carried out to ensure a numerically stable trend solution from all data that can be evaluated in the test area.

By their nature, magnetic field data are often vectorial and/or tensorial and as such can be evaluated linearly, unlike gravimetrically/magnetometrically recorded intensities (root magnitudes), which are nonlinear in character. Due to the complex source origin of the magnetic field ingredients, the data field processing and modeling are by far more complex than in gravimetry. Moreover, in contrast to the gravitational field, the time variability in the magnetic field data has to be taken into account, especially in the case of temporally heterogeneous data.

3.2 Basic modeling in terms of Fredholm integral equations

Not only gravimetry and magnetometry (see also [34, 54, 71] for survey documentations) lead to potential methodologies, although the measuring instrumentarium of other methods is different from those we know for gravimetry and magnetometry. As a matter of fact, as we have seen, similar mollifier techniques as deduced for gravimetry and magnetometry can be developed for other problems in geomathematics, for example, acoustically and elastically based scattering (cf. [28, 29, 30]). Moreover, poroelastic modeling (cf. [9, 55, 113] for some entry studies) plays a particular role in the near future.

In a mathematical abstraction (see, e. g. [60]), the gravimetric and magnetometric approaches for an area $\mathcal{B}$ of interest inside the Earth can be identified as prototypes for some other potential methods in geothermal research, which may be characterized as Fredholm integral equation of the first kind in a condensed formulation as follows:

$$V(x) = \int_{\mathcal{B}} G(L;x,y)\ F(y)\ dV(y) \tag{3.1}$$

involving the fundamental solution $G(L;x,y)$ corresponding to a differential operator L so that

$$F(x) = LV(x), \quad x \in \mathcal{B}, \tag{3.2}$$

where (in distributional sense) by virtue of the δ-distribution

$$L_x G(L;x,y) = \delta(x,y), \quad x \in \mathcal{B}, \tag{3.3}$$

and

$$F(x) = \int_{\mathcal{B}} \delta(x,y)\ F(y)\ dV(y), \quad x \in \mathcal{B} \tag{3.4}$$

(dV is the volume element).

In specific dependence on the choice of the differential operator L we are led to the following exploration areas:

(i) *gravitational modeling* (L (monopole-based negative) Laplace operator),
(ii) *geomagnetic modeling* (L (dipole-oriented) pre-Maxwell operators),
(iii) *acoustic seismic tomography* (L (negative) Helmholtz operator),
(iv) *elastic seismic tomography* (L Cauchy-Navier operator),
(v) *acoustic scattering* (L time dependent acoustic wave equation),
(vi) *elastic scattering* (L time dependent elastic wave equation).

Finally, from the point of view of mathematics, it should be mentioned that decorrelation methods as presented here were applied in the following publications concerned with exploration techniques:

(i) gravity modeling (cf. [31, 33, 34, 53, 54, 55, 56, 57, 63, 77]),
(ii) geomagnetic field modeling (cf. [34, 54, 57, 86]),
(iii) acoustic seismic (cf. [28, 34, 54, 56, 60]),
(iv) elastic seismic (cf. [29, 30]),
(v) poroelasticity and stress behavior (cf. [8, 9, 10, 55, 71, 113]).

The geomathematical treatment of tomography and scattering problems with time-dependent acoustic or elastic wave operators in a consistent context that provides further deeper information about Earth science subsystems is a promising challenge for future research.

3.3 Monopole gravimetric modeling

Basically, two types of inverse problems can be distinguished for geologic exploration:

(i) Calculation of the density distribution within an area $\mathcal{B}$ of the Earth's interior (and thus of information about geologic strata in this area) from measured gravitational field information on the Earth's surface.

(ii) Calculation of gravitational field information for an area $\mathcal{B}$ of the Earth's surface from gravitational field information measured at satellite height (or aircraft altitude).

A combination of both problems, i. e., the calculation of density information from gravitational field information in the Earth's outer space, is also possible and mathematically senseful, but combined methodologies are hardly realizable because of the spectral exponential decay of satellite data (cf. [74]) in computational practice. Nevertheless, as already explained, data of the Earth's outer space are helpful in suppressing undesirable oscillatory (Gibb's) phenomena.

Mathematically, both types (i) and (ii) of calculations have in common that they lead to a Fredholm integral formula of the first type in the form

$$\langle K(x,\cdot),\ F\rangle_{L^2(\mathcal{B})} = V(x), \quad x \in \overline{\mathcal{B}}, \tag{3.5}$$

with a given kernel function $K : \overline{\mathcal{B}} \times \overline{\mathcal{B}} \to \mathbb{R}$. $\mathcal{B}$ is the corresponding integration area under investigation, and $\overline{\mathcal{B}} = \mathcal{B} \cup \partial\mathcal{B}$ is the topological closure of $\mathcal{B}$, where by convention $\partial\mathcal{B}$ denotes the boundary of $\mathcal{B}$. Characteristic is the inverse character of the integral equation (3.5), namely that it aims at a measured "effect" V to a desired "cause" F. $V(x)$ is provided by observation discretely for points x in the "measurement range". Ideally, F is sought continuously for all points of the "range of causes".

In the following, we will describe the three essential problems we are confronted with in inverse problems in accordance with the considerations already made earlier:

(i) *Inverse gravimetry by "downward computation"* following Newton's law is given by a volumetric integral equation, which is a convolution of the density distribution F with the Newton kernel, i. e., the reciprocal distance between the integration variable and the computation point.

(ii) *Inverse gradiometry by "downward continuation"* leads, under the assumption that satellite orbit and Earth's surface are each related to spherical reference surfaces according to Runge's approach (see [11, 50, 52, 70, 71]), to a spherical integral formula which has as kernel certain second derivatives of the so-called Poisson kernel.

(iii) *Inverse gradiometry/gravimetry by "downward computation"* is either a composition of (i) and (ii) or it canonically results from a modification of the Newtonian volume potential.

Classical gravimetry. As already mentioned, the point of departure in the case of classical gravimetry is Poisson's differential equation, which is using the Laplace operator Δ for a geometrically specified domain $\mathcal{B} \subset \mathbb{R}^3$, whose boundary $\partial\mathcal{B}$ allows the application of the theorem of Gauss known from vector analysis by

$$\rho = -\Delta V \tag{3.6}$$

where, according to the Newtonian theory, V is the gravitational potential to the sufficiently smooth (e. g., Hölder continuous) density distribution F

in $\overline{\mathcal{B}}$, which is conventionally designated by ρ in the geoscientific literature. The fundamental solution of the Laplace operator Δ, i. e., the (monopole) Newton kernel, is given by the rotationally symmetric kernel function

$$K(x,y) = G(x,y) \tag{3.7}$$

with

$$G(x,y) = \frac{1}{4\pi}\frac{1}{|x-y|}, \quad x,y \in \mathbb{R}^3,\ x \neq y \tag{3.8}$$

(see e. g. [31, 56, 57] for further potential-theoretic explanations). The general solution of the Poisson equation (3.6) can be calculated with the help of the fundamental solution of the Laplace operator Δ in the form of the (volumetric) Newtonian integral representation

$$V(x) = \int_{\mathcal{B}} \underbrace{\frac{1}{4\pi}\frac{1}{|x-y|}}_{\text{monopole kernel}} \rho(y)\, dV(y), \quad x \in \mathbb{R}^3, \tag{3.9}$$

where dV in the case of a volumetric domain $\mathcal{B} \subset \mathbb{R}^3$ is the volume element in $\mathbb{R}^3$.

Already at this early stage it should be pointed out that the interest of this work is simply laid in studying the gravitational field in macroscopic sense, where the quantum behavior of gravitation may not be taken in account. Moreover, our stringent, but not always physically tolerable assumption is that Newtonian physics can be safely used for our numerical realizations (note that the numerics of more general approaches is a challenge for future work, which will not be discussed here).

Once again, the classical application in geoscience results from the so-called *inverse gravimetry problem*, i. e., the determination of the density function ρ in $\overline{\mathcal{B}}$ appearing in the Newton integral (3.9) from information of the gravitational potential P, usually given on the boundary $\partial\mathcal{B}$ of the domain, but also from the outer space (e. g., in case of measurement with flying objects with low flight altitude).

Observable-based gravimetry. With this method instead of the potential values $P(x)$ for points $x \in \partial\mathcal{B}$ (or more generally also in $\mathbb{R}^3 \backslash \mathcal{B}$) other geodetic measurements (such as gravitational anomalies) are used. The methodical procedure is in this case largely analogous.

Classical gradiometry. As usually we denote the sphere around zero with radius R as $\mathbb{S}^2_R$, i. e., $\mathbb{S}^2_R = \{x \in \mathbb{R}^3 : |x| = R\}$. We further assume for simplicity that the Earth's surface $\mathbb{S}^2_R$ and the satellite's orbit $\mathbb{S}^2_r$, $R < r$, are spheres around zero (center of gravity of the spherical Earth model) with radii R and

r. The altitude of the satellite is given by $r - R$. The ball around the origin with radius R is defined by $\mathbb{B}_R^3 = \{x \in \mathbb{R}^3 : |x| < R\}$.

Upward continuation. The so-called "upward continuation" assigns a function F on the sphere $\mathbb{S}_R^2$ (Earth's surface) the corresponding values in the outer space (e. g., on the satellite orbit). Mathematically formulated, we consider the mapping A that assigns the function V to a function $F \in L^2(\mathbb{S}_R^2)$, i. e., $A : F \to V = A[F]$ with

$$K(x,y) = \frac{1}{4\pi R} \frac{|x|^2 - R^2}{|x-y|^2}, \tag{3.10}$$

and

$$V(x) = A[F](x) = \int_{\mathbb{S}_R^2} \underbrace{\frac{1}{4\pi R} \frac{|x|^2 - R^2}{|x-y|^2}}_{\text{Poisson kernel}} F(y)\, dS(y), \quad |x| > R, \tag{3.11}$$

where dS denotes the surface element. The result of this mapping is a harmonic function V in the outer space of the sphere $\mathbb{B}_R^3$, i. e., in $\mathbb{R}^3 \backslash \overline{\mathbb{B}_R^3}$, i. e.,

$$x \mapsto V(x) = A[F](x), \quad x \in \mathbb{R}^3 \backslash \overline{\mathbb{B}_R^3}, \tag{3.12}$$

satisfies Laplace's equation

$$\Delta V(x) = 0, \quad x \in \mathbb{R}^3 \backslash \overline{\mathbb{B}_R^3}, \tag{3.13}$$

in the outer space. The space $A[L^2(\mathbb{S}_R^2)]$ forms a Hilbert space of Sobolev type (see [52]). For any $r > R$ we have that $A[L^2(\mathbb{S}_R^2)]|_{\mathbb{S}_r^2}$ is a real subspace of $L^2(\mathbb{S}_r^2)$.

Downward continuation. The inverse problem of "upward continuation" is to determine the square-integrable function F on the Earth's surface $\mathbb{S}_R^2$ from knowledge of the values of the potential V on the satellite orbit $\mathbb{S}_r^2$, i. e.,

$$V(x) = A[F](x), \quad x \in \mathbb{S}_r^2. \tag{3.14}$$

Since, as mentioned above, $A[L^2(\mathbb{S}_R^2)]|_{\mathbb{S}_r^2}$ is a real subspace of $L^2(\mathbb{S}_r^2)$, the problem of "downward continuation" is not solvable for all functions in $L^2(\mathbb{S}_r^2)$. "Downward continuation" is only feasible, if the integral equation $A[F] = V$ has a right side which is a restriction of a potential V on $\mathbb{S}_r^2$. This issue, however, is not of practical relevance, since to model F one can assume discretely given values of V on the satellite orbit $\mathbb{S}_r^2$.

For modeling of essential importance is the following fact: The operator A represents a bounded mapping of the space $L^2(\mathbb{S}_R^2)$ into the Sobolev space $A[L^2(\mathbb{S}^2_R)]|_{\mathbb{S}^2_r} \subset L^2(\mathbb{S}^2_r)$. It even has a uniquely defined inverse on $A[L^2(\mathbb{S}^2_R)]|_{\mathbb{S}^2_r}$. Unfortunately, according to the laws of functional analysis, this inverse is unbounded on the space $A[L^2(\mathbb{S}^2_R)]|_{\mathbb{S}^2_r}$, i. e., the inversion process cannot be carried out stably. It requires a so-called "regularization" by approximating the unbounded inverse operator by a family of bounded operators. In the modeling setup, the unbounded operator is replaced by a bounded regularization operator.

The assumption of a spherical satellite orbit is not mandatory for the above described problem. "Downward continuation" can thus also be understood more generally as the inverse problem of determining $F \in L^2(\mathbb{S}^2_R)$ from the assumption of a potential V of the Sobolev space $A[L^2(\mathbb{S}^2_R)]$ (entire outer space of the Earth) instead of $A[L^2(\mathbb{S}^2_R)]|_{\mathbb{S}^2_r}$ (spherical satellite orbit). As a consequence, discrete potential values from all over the outer space of the Earth can find their way into the inverse problem. The uniquely determined inverse remains of course unbounded.

3.4 Dipole magnetometric modeling

Magnetometry as an inverse problem formally understood in analogy to gravimetry means the determination of the vectorial magnetization m from the magnetic field potential B produced by susceptible material in $\overline{\mathcal{B}}$ based on a scalar Fredholm integral equation

$$B(x) = \int_{\mathcal{B}} \underbrace{K(x,y)}_{\text{dipole kernel}} \cdot m(y)\, dV(y) \tag{3.15}$$

with

$$K(x,y) = \gamma(x,y) \tag{3.16}$$

and

$$\gamma(x,y) = \nabla_y G(x,y) = \nabla_y \frac{1}{4\pi} \frac{1}{|x-y|}, \quad x,y \in \mathbb{R}^3,\ x \neq y. \tag{3.17}$$

Inverse magnetometry understood in this formal sense shows similar difficulties as the gravitational case via the Newton integral equation.

The magnetostatic volume integral (3.15) over $\mathcal{B}$ corresponding to a vectorial magnetization distribution m fulfills the Laplace equation in the external space $\mathcal{B}^c = \mathbb{R}^3 \backslash \overline{\mathcal{B}}$ ($\overline{\mathcal{B}} = \mathcal{B} \cup \partial\mathcal{B}$, $\partial\mathcal{B}$ boundary of $\mathcal{B}$). This property is a direct consequence of the harmonicity of the monopole Newtonian kernel.

The goal of inverse magnetometry is to determine the magnetization distribution m within a subarea $\mathcal{B}$ of the Earth. Usually, one assumes a vector field of limited signal energy in $\overline{\mathcal{B}}$, i. e.,

$$\|m\|_{l^2(\mathcal{B})} = \left(\int_{\mathcal{B}} |m(x)|^2 \, dV(x) \right)^{1/2} < \infty. \tag{3.18}$$

Under this mathematical condition, the *magnetic dipole potential*

$$B(x) = \frac{1}{4\pi} \int_{\mathcal{B}} \nabla_y \frac{1}{|x-y|} \cdot m(y) \, dV(y), \quad x \in \mathbb{R}^3 \tag{3.19}$$

and the *dipole magnetic induction*

$$b(x) = \frac{1}{4\pi} \nabla_x \int_{\mathcal{B}} \nabla_y \frac{1}{|x-y|} \cdot m(y) \, dV(y), \quad x \in \mathbb{R}^3, \tag{3.20}$$

can be calculated anywhere in $\mathbb{R}^3$, so that the *direct magnetometry problem*

$$\underbrace{m}_{=\text{ magnetization distribution}} \longrightarrow \underbrace{B}_{=\text{ magnetic dipole potential}} \tag{3.21}$$

can be realized with the help of approximate integration.

As far as the point x is located in the outer space $\mathcal{B}^c = \mathbb{R}^3 \backslash \overline{\mathcal{B}}$, $B(x)$ is obtainable as regular integral. However, if the point x is located in the interior $\mathcal{B}$ or on the boundary $\partial\mathcal{B}$, we are confronted with an improper integral. Therefore, analogous to the gravimetric problem, one can assume that inverse magnetometry

$$\underbrace{B}_{=\text{ magnetic dipole potential}} \longrightarrow \underbrace{m}_{=\text{ magnetization distribution}} \tag{3.22}$$

also shows a specific behaviour depending on the position of the input data.

In operator formulation the integral equation (3.19) is given in the form

$$A[m] = B, \tag{3.23}$$

where the operator A can be written as volume integral over $\mathcal{B} \subset \mathbb{R}^3$

$$A[m](x) = \int_{\mathcal{B}} \nabla_y G(x,y) \cdot m(y) \, dV(y) = B(x), \quad x \in \mathbb{R}^3. \tag{3.24}$$

An investigation of the ill-posedness in the sense of Hadamard is similar to that of inverse gravimetry. It can be found, e. g., in [115, 153] (see also the references therein).

3.5 Conceptional comparison of potential methods

Inverse problems, which strive to infer from the effect to its cause, are usually ill-posed in the sense that at least one of the three properties existence, uniqueness, stability is violated. For inverse gravimetry and magnetometry it is known (cf. e. g. [13, 27, 115, 123, 124, 142, 153, 177]) that even all three properties are violated for terrestrial data specifications. The procedure for resolution ("approximate solvability") of ill-posed inverse problems consists in constructing a "solution" that is physically acceptable, sufficiently close in approximation, and sufficiently stable in numerics. Ill-posedness thus forces a distinction between "solution" and "resolution" (see, e. g., [62, 66] for more details). In this sense, the mollifier reconstruction implemented in this work is a realistic resolution method.

For the convenience of the reader we summarize the basic equations of inverse gravimetry and magnetometry (already presented earlier in this chapter): Gravity as well as magnetic field inversion for the determination of the density distribution or material magnetization in the subsurface can be described by inverse problems of the potential theory. Both problems can be mathematically captured by a volume singular Fredholm integral equation of the first kind. Using well-known potential-theoretic approaches (see, e. g., [27, 34, 54]) from the *Newton's law of gravitation* it follows the scalar integral equation

$$V(x) = \int_{\mathcal{B}} G(x,y)\ \rho(y)\ dV(y), \quad x \in \mathbb{R}^3, \tag{3.25}$$

for the *gravitational potential* V *from the density distribution* F *in* $\mathcal{B}$ and the vectorial integral equation

$$v(x) = \nabla_x \int_{\mathcal{B}} G(x,y)\ \rho(y)\ dV(y), \quad x \in \mathbb{R}^3, \tag{3.26}$$

for the *gravitational field* v *from the density distribution* F *in* $\mathcal{B}$ as well as from the *Biot-Savart law* the scalar integral equation

$$B(x) = \int_{\mathcal{B}} \gamma(x,y) \cdot m(y)\ dV(y), \quad x \in \mathbb{R}^3, \tag{3.27}$$

for the *magnetic potential* B *from the magnetization distribution* m *in* $\mathcal{B}$ and the vector integral equation

$$b(x) = \nabla_x \int_{\mathcal{B}} \gamma(x,y) \cdot m(y)\ dV(y), \quad x \in \mathbb{R}^3, \tag{3.28}$$

for the *magnetic vector field b from the magnetization distribution m in $\mathcal{B}$.*

It should be remarked that the gravitational constant as well as the vacuum permeability are set to one in the formulas for simplicity. In the numerical calculations later, however, they are considered accordingly.

In the case of gravimetry, the equations (3.25), (3.26) define a scalar source kernel function (monopole kernel)

$$(x,y) \mapsto G(x,y) = \frac{1}{4\pi|x-y|}, \quad x \neq y, \tag{3.29}$$

while, in the case of magnetometry, (3.27), (3.28) is predefined by a vector source kernel function (dipole kernel)

$$(x,y) \mapsto \gamma(x,y) = \nabla_y G(x,y) = -\frac{x-y}{4\pi|x-y|^3}, \quad x \neq y \tag{3.30}$$

(cf. also Table 3.1). According to their source character, both kernel functions are singular at the position $x = y$. However, according to the mathematical classification (cf., e. g., [66, 72]), the singularity is to be classified as weak in the context of the theory of singular integral equations. An elementary calculation shows for $x \neq y$ the relation that with respect to the variable y the scalar source kernel function is the gradient field of the vector source kernel function and with respect to the variable x the scalar source kernel function is the negative gradient field of the vector source kernel function:

$$\begin{aligned}\gamma(x,y) &= \nabla_y G(x,y) = \nabla_y \frac{1}{4\pi|x-y|} \\ &= -\nabla_x G(x,y) = -\frac{x-y}{4\pi|x-y|^3}.\end{aligned}$$

Exploration method	**Gravimetry**	**Magnetometry**
Physical context	Newton's law of gravity	Biot-Savart law
Input variable	Gravitational potential V	Magnetic induction b
Output variable	Density ρ	Magnetization m
Underlying differential equation	Scalar Laplace equation	Vectorial (Pre-)Maxwell equation
Field type	Monopole field	Dipole field
Source kernel function	$G(x,y)$	$\gamma(x,y)$

Table 3.1 Comparison of gravimetry and magnetometry.

In the sense of Hadamard's classification commonly used in mathematics (see, e. g., [72]), we know that the singular integral equations (3.25), (3.26), resp. (3.27), (3.28) lead to inverse problems for the determination of F and

m, respectively, which are not well-posed in terms of existence, uniqueness, and numerical stability, the nonuniqueness having large implications for the data to be assumed.

As we already mentioned earlier, in the center of the mathematical solution context of the integral equations, there is formally a physically motivated equation

$$-\Delta_x G(x,y) = \delta(x,y), \tag{3.31}$$

which shows the negative Laplace derivative of the source kernel function $G(\cdot,\cdot)$ as Diracian delta function(s) $\delta(\cdot,\cdot)$. This then leads mathematically to the so-called *Poisson's differential equation*

$$-\Delta V(x) = \int_{\mathcal{B}} (-\Delta_x) G(x,y)\ \rho(y)\ dV(y) = \int_{\mathcal{B}} \delta(x,y)\ \rho(y)\ dV(y) = \rho(x) \tag{3.32}$$

at least under certain smoothness conditions known from potential theory at ρ (see, e. g., [57]), but only for points x of the considered area $\mathcal{B}$.

3.6 Novel methodologies and expectable potential signatures

Our purpose is to summarize the results obtained until now: During the last three decades, the rapid development in computer technology and the enormous increase in measurement and observation methods have led to an explosive spread of new mathematical approaches, such as wavelet methods. Mathematics as an interdisciplinary science can be found in almost every area of our lives ("mathematization of science"). It acts as a key technology in all branches of science, precisely also in the field of exploration. In the form developed by the authors, it allows modeling, visualization and simulation even of complicated structures and processes while processing huge (noisy) amounts of data on a previously unknown scale. The potential quantities measured in gravimetric or magnetometric exploration procedures are integrated effects of the subsurface and have to be decorrelated due to their superposition character to ensure geologic interpretability and applicability in geothermal energy. As the ultimate outcome for our intentions in geologic exploration it may be desirable that the transition surfaces characterizing geologic formations and manifesting themselves in their phenotype by high-frequency signatures can be increasingly better prepared for the contrast function.

In continuation of GEOFÜND (PtJ FKZ 0325512A), the projects SPE (PtJ FKZ 0324061) and SYSEXPL (PtJ FKZ 03EE4002A) dealt with novel geomathematical evaluation mechanisms with the following intentions, which should be explained, tested, and documented in more detail in the remaining part of this work:

- Geologic interpretation by an appropriate processing of available potentials and contrast functions also in comparison of different solution proposals by decorrelative exploration,
- detection of geologic formations and discontinuities as well as localization of, e. g., aquifers based on preliminary geologic information (before and during the drilling phase),
- local inversion of the underlying integral equation for the calculation of the contrast function on a subarea of the Earth relevant for the borehole with additional use of borehole data, i. e., functionals of the potential (prior information and/or knowledge after insertion of the first borehole).

So, the essential concepts of the book will demonstrate the following two aspects:

- Two independent and thus mutually validatable inverse decorrelation methods for determining the subsurface from gravimetric and magnetometric measurements are newly developable and applicable for geothermal exploration. These methods allow to realistically decompose the gravitational or magnetic field signatures into so-called bandpass information and thus to obtain an improved understanding of the subsurface geology by detailed information on high-frequency signatures. Indeed, in approval of [34, 54], the presented multiscale methods open up new evaluation aspects in the decorrelation of geophysical data by locally based interpretation methods. The fine structure modeling of the subsurface is thereby based on the locally available data. The procedure ensures modeling as close to the physics as feasible, allowing interpretation as close to reality as possible through designation of decorrelated parting planes, and thus a better understanding of the geologic formations ideally obtained from the synopsis of gravimetric and magnetometric results. A meaningful interpretation of the results offers a deeper understanding of the geologic structures.

- A spatially-dependent complementary system considering prior information, data type, data accuracy and data density as well as modeling specifics and results in functional interrelations, which, as already mentioned, is urgently needed especially for investors without a geothermal or mining background, in order to also make geothermal energy more applicable in densely built-up areas or with anthropogenically pre-polluted subsurface. As outcome we will be able to state that modern multiscale methods have a lot to offer by their ability to decorrelate potential and wave signatures spatially and temporally into suitable band structures and to specify geothermally relevant geologic information via a mathematical inversion process. In addition, new measurement and observation methods (such as sensitive gravity gravimetry by atomic interferometry or tensorial gradiometer measurements in magnetics) will open up much higher measurement accuracies and thus improved modeling possibilities in the near future.

Chapter 4

Gravimetric multiscale approximation and mollifier decorrelation

Abstract:

The specific idea of the mollifier method in the theory of inverse problems is a "replacement (re)solution", where one finds a too "strong ill-posedness" to bring one of the standard regularization methods (e. g. the computation of a pseudoinverse) into use. A heuristic motivation for a "mollification" is on the one hand the annoyance that high-frequency components in the solution have to be specifically damped by certain smoother structures in order to arrive at (re)solution statements at all. On the other hand, a certain amount of high-frequency components are needed to tease out transition structures in geologic signatures for decorrelation. So, at the end of a mollifier regularization, a compromise should be able to provide a wavelet decorrelation of geologic strata as much as possible in terms of geophysically relevant approximation structures close to reality, however, under (often disadvantageous) constraints of the data situation and the measurement accuracy.

4.1 Concept of mollifier regularization

In the following, we draw on the approach [56] proposed in the journal "World of Mining" from 2013. Our goal is to describe mollifier decorrelation of geologic signatures by potential-theoretic multiscale approximation. First, we specifically use gravimetry as a prototype. Completely analogous mathematical procedures apply to magnetometry, which are immediately obvious from

W. Freeden, H. Nutz, *Exploratory Potential Methods in Geothermal Power Generation*, Lecture Notes in Geosystems Mathematics and Computing,
https://doi.org/10.1007/978-3-031-54412-5_4

the exposition and do not require a detailed recapitulation (cf. also [34, 54]). The most important tools turn out to be mollifier kernel structures to the singular source kernel functions $G(\cdot,\cdot)$ as well as $\gamma(\cdot,\cdot)$, which allow the introduction of corresponding scale-dependent regular mollifier kernels and thus, because of their mollifier character, allow a geologically evaluable multiscale access via regular integral equations.

Mollifiers were originally introduced by K.O. Friedrichs in a 1944 paper [81], which is considered one of the turning points in the modern theory of partial differential equations. In a somewhat extended understanding, mollifiers today represent auxiliary functions, e. g., in constructive approximation, to generate sequences of smooth (regular) functions that approximate a nonsmooth (usually singular) function (such as $G(\cdot,\cdot)$ or $\gamma(\cdot,\cdot)$). Such smoothing ("mollification") makes the sharp features of an irregular signature "softer" ("mollified"), but keeps them close to the original nonsmooth signature, which is of enormous practical importance for realistic solution modeling (cf. [54, 34]).

Specifically for our exploration applications the following characterization is important: A mollifier is a (scale-dependent) regular kernel function that is convolved with the contrast function (for example, the Earth's internal density distribution, Earth's internal magnetization distribution), yielding a (scale-dependent) filtered function that is close to, but smoother than, the contrast function. In other words, in the approximation to a contrast function, a mollifier is not just a kernel function, but a family or, more precisely, a one-parameter (i. e., scale-dependent) sequence of kernel functions (see Figure 4.1). For purposes of decorrelation, the difference of two subsequent scale-dependent filtered functions is simply needed to obtain the information included in the filtered function with the higher scale, which is not included in the filtered function with the lower scale. In continuing so with increasing scales, we finally obtain interpretable knowledge of transitional structures and/or discontinuities in the contrast function, as far as the chosen difference (wavelet) mollifiers are geophysically relevant.

4.2 Concept of gravimetric mollifier regularization

All the subsequent results are now based on the following statement, which in our case is directly formulated in a scale-discrete context ready for numerics:

Zero sequence. We consider a monotonically decreasing zero sequence $\{\tau_j\}_{j\in\mathbb{N}_0}$, i. e., $\tau_{j+1} \le \tau_j$, $j \in \mathbb{N}_0$, and

$$\lim_{j\to\infty} \tau_j = 0. \tag{4.1}$$

To each monotonically decreasing zero sequence $\{\tau_j\}_{j\in\mathbb{N}_0}$ there exists a sequence of (regular) "mollifier kernels"

$$(x,y) \mapsto G_{\tau_j}(x,y), \quad x,y \in \mathbb{R}^3, \tag{4.2}$$

with the property of striving toward the singular kernel

$$\lim_{j\to\infty} G_{\tau_j}(x,y) = G(x,y), \quad x,y \in \mathbb{R}^3, \; x \neq y, \tag{4.3}$$

and satisfying the limit relation

$$\lim_{j\to\infty} G_{\tau_j}(x,y) = G(x,y), \quad x,y \in \mathbb{R}^3, \; x \neq y, \tag{4.4}$$

such that the τ_j*-mollifier potential functions* V_{τ_j} given by

$$V_{\tau_j}(x) = \int_{\mathcal{B}} G_{\tau_j}(x,y)\; \rho(y)\; dV(y), \quad x \in \mathbb{R}^3, \tag{4.5}$$

and the τ_j*-mollifier contrast functions* ρ_{τ_j} given by

$$\rho_{\tau_j}(x) = \int_{\mathcal{B}} H_{\tau_j}(x,y)\; \rho(y)\; dV(y), \quad x \in \mathbb{R}^3, \tag{4.6}$$

are characterized by the limit relation

$$\begin{aligned} -\Delta V(x) &= \lim_{j\to\infty} -\Delta V_{\tau_j}(x) \\ &= \lim_{j\to\infty} \int_{\mathcal{B}} (-\Delta_x) G_{\tau_j}(x,y)\, \rho(y)\; dV(y) \\ &= \lim_{j\to\infty} \int_{\mathcal{B}} H_{\tau_j}(x,y)\; \rho(y)\; dV(y) \\ &= \lim_{j\to\infty} \rho_{\tau_j}(x) \\ &= \rho(x). \end{aligned} \tag{4.7}$$

The kernels $G_{\tau_j}(\cdot,\cdot)$ and $H_{\tau_j}(\cdot,\cdot)$, $j \in \mathbb{N}_0$, are called Dirac's scaling functions (sometimes also Green's scaling functions) to scale $j \in \mathbb{N}_0$ (see Figure 4.1).

The difference kernels

$$\Psi_{G_{\tau_j}}(x,y) = G_{\tau_{j+1}}(x,y) - G_{\tau_j}(x,y), \quad x,y \in \mathbb{R}^3, \tag{4.8}$$

and

$$\Psi_{H_{\tau_j}}(x,y) = H_{\tau_{j+1}}(x,y) - H_{\tau_j}(x,y), \quad x,y \in \mathbb{R}^3, \tag{4.9}$$

are called Dirac wavelet functions (sometimes also Green wavelet functions) to scale $j \in \mathbb{N}_0$ (see Figure 4.2).

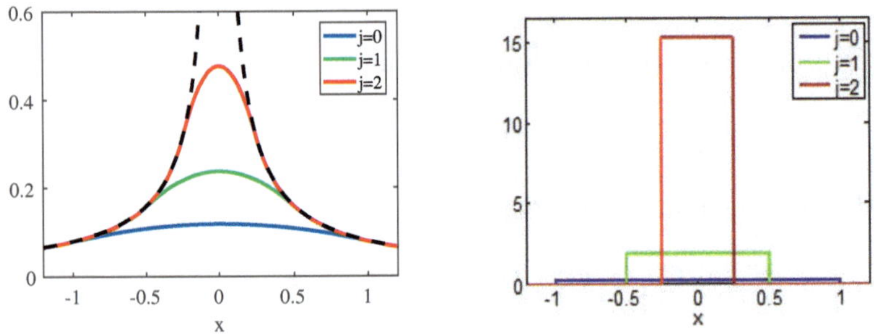

Fig. 4.1 Profiles of the scaling functions $G_{\tau_j}(\cdot,\cdot)$ (left) and $H_{\tau_j}(\cdot,\cdot)$ (right), respectively, for the dyadic choice $\tau_j = 2^{-j}$. (H_{τ_j} is a Dirac scaling function of Haar type, other examples can be found in [54]).

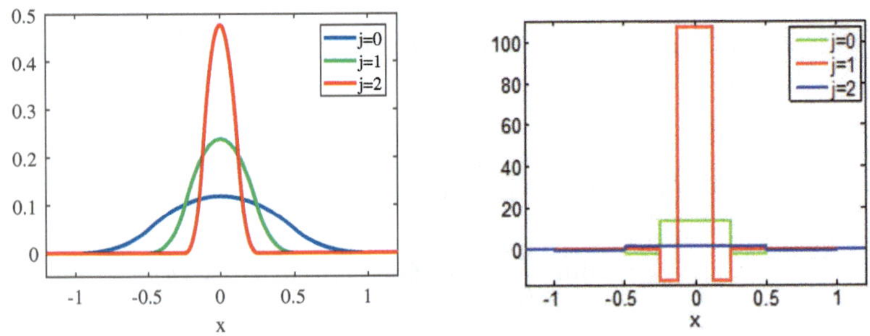

Fig. 4.2 Profiles of the wavelet functions $\Psi_{G_{\tau_j}}(\cdot,\cdot)$ (left) and $\Psi_{H_{\tau_j}}(\cdot,\cdot)$ (right) for the choice $\tau_j = 2^{-j}$, respectively.

The τ_j wavelet potential functions

$$(WV)_{\tau_j}(x) = \int_{\mathcal{B}} \Psi_{G_{\tau_j}}(x,y)\ \rho(y)\ dV(y), \quad x \in \mathbb{R}^3, \tag{4.10}$$

and the τ_j wavelet contrast functions

$$(W\rho)_{\tau_j}(x) = \int_{\mathcal{B}} \Psi_{H_{\tau_j}}(x,y)\ \rho(y)\ dV(y), \quad x \in \mathbb{R}^3, \tag{4.11}$$

characterize successive decorrelations, i. e., detailed information ("band structures") contained in the datasets defined by the differences $V_{\tau_{j+1}} - V_{\tau_j}$ or $\rho_{\tau_{j+1}} - \rho_{\tau_j}$ are included. More precisely, the "band structure" is denoted by

$$(WV)_{\tau_j}(x) = V_{\tau_{j+1}}(x) - V_{\tau_j}(x), \quad x \in \mathbb{R}^3, \tag{4.12}$$

as well as

$$(W\rho)_{\tau_j}(x) = \rho_{\tau_{j+1}}(x) - \rho_{\tau_j}(x), \quad x \in \mathbb{R}^3, \tag{4.13}$$

i. e., $(WV)_{\tau_j}$ or $(W\rho)_{\tau_j}$ is the mollifier detail information contained in $V_{\tau_{j+1}}$ or $\rho_{\tau_{j+1}}$, respectively, but not in V_{τ_j} or ρ_{τ_j}.

Of particular importance from a numerical point of view is the special construction of the wavelet functions $\Psi_{G_{\tau_j}}(\cdot,\cdot)$, $\{\tau_j\}_{j\in\mathbb{N}_0}$. They are defined (cf. the graphical illustrations in Figures 4.1 and 4.2) such that the functions $\Psi_{G_{\tau_j}}(\cdot,\cdot)$ have a local support, i. e., as functions of variables y they vanish outside a sphere $\mathcal{B}_{\tau_j}(x)$ of radius $\tau_j > 0$ with center x. Expressed in formulas this means

$$(WV)_{\tau_j}(x) = V_{\tau_{j+1}}(x) - V_{\tau_j}(x) = 0, \quad x \in \mathcal{B}\backslash\overline{\mathcal{B}_{\tau_j}(x)}. \tag{4.14}$$

Our approach initiated in the paper [56] and explained above for the gravimetric case therefore implies that the integral is not to be extended over the full integration domain $\mathcal{B}$, but has a local spherical support:

$$(WV)_{\tau_j}(x) = \int_{\mathcal{B}_{\tau_j}(x)\cap\mathcal{B}} \Psi_{G_{\tau_j}}(x,y)\ \rho(y)\ dV(y), \quad x \in \mathcal{B}. \tag{4.15}$$

As j becomes larger, the sphere supports become smaller and smaller. For each $J \in \mathbb{N}$ the telescope sums

$$\sum_{j=0}^{J-1}(WV)_{\tau_j}(x) = \sum_{j=0}^{J-1}\left(V_{\tau_{j+1}}(x) - V_{\tau_j}(x)\right) \tag{4.16}$$

and

$$\sum_{j=0}^{J-1}(W\rho)_{\tau_j}(x) = \sum_{j=0}^{J-1}\left(\rho_{\tau_{j+1}}(x) - \rho_{\tau_j}(x)\right) \tag{4.17}$$

immediately provide the identities

$$V_{\tau_J}(x) = V_{\tau_0}(x) + \sum_{j=0}^{J-1}(WV)_{\tau_j}(x) \tag{4.18}$$

and

$$\rho_{\tau_J}(x) = \rho_{\tau_0}(x) + \sum_{j=0}^{J-1} (W\rho)_{\tau_j}(x), \tag{4.19}$$

such that at the boundary transition $J \to \infty$ the regular mollifier potential functions V_{τ_J} tend towards the singular potential V

$$V(x) = \lim_{J\to\infty} V_{\tau_J}(x) = V_{\tau_0}(x) + \sum_{j=0}^{\infty} (WV)_{\tau_j}(x) \tag{4.20}$$

and simultaneously the mollifier contrast functions ρ_{τ_J} tend towards the density ρ

$$\begin{aligned} \rho(x) &= \lim_{J\to\infty} \rho_{\tau_J}(x) = \rho_{\tau_0}(x) + \sum_{j=0}^{\infty} (W\rho)_{\tau_j}(x) \\ &= \lim_{J\to\infty} -\Delta V_{\tau_J}(x) = -\Delta V_{\tau_0}(x) + \sum_{j=0}^{\infty} (-\Delta)(WV)_{\tau_j}(x). \end{aligned} \tag{4.21}$$

The formulas (4.20) and (4.21) indicate the multiscale characteristic, i. e., that $V(x)$ and $\rho(x)$, respectively, result from a lowpass filtering $V_{\tau_0}(x)$ and $\rho_{\tau_0}(x)$ and successive addition of bandpass filtering $(WV)_{\tau_j}(x)$ and $(W\rho)_{\tau_j}(x)$, respectively, $j = 0,1,\ldots$. As the scale parameter j increases, $(WV)_{\tau_j}$ as well as $(W\rho)_{\tau_j}$ provide refining detail information, which in a geologic scenario implies increasingly emerging high-frequency signature structures, i. e., transition surfaces. This applies equally to all mollifier wavelets of gravimetry and magnetometry (as well as to the exploration techniques of acoustic and elastic seismics addressed in the PtJ project GEOFÜND (PtJ FKZ 0325512A) cf. [28, 34, 53, 54, 56]). Particularly significant from a numerical point of view is that the locally compact supports $\overline{\mathcal{B}_{\tau_j}(x)}$ have a decreasing radius τ_j as the scale parameter j increases. This allows locally an increase of the data density simultaneously with the spatial localization of the supports, if this is feasible from a metrological point of view. Thus, data collection for wavelet decorrelation should be increased where geologic modeling features and specifics are expected.

4.3 Gravimetric mollifier functions

For a monotonically decreasing zero sequence $\{\tau_j\}_{j\in\mathbb{N}_0}$, e. g., the dyadic sequence $\tau_j = 2^{-j}$, $j \in \mathbb{N}_0$, we can use a linear Taylor approximation (cf. [54] for more details) to find the kernel functions

$$G_{\tau_j}(x,y) = \begin{cases} \frac{1}{8\pi\tau_j}\left(3 - \frac{1}{\tau_j^2}|x-y|^2\right), & |x-y| \le \tau_j, \\ \frac{1}{4\pi|x-y|}, & |x-y| \ge \tau_j, \end{cases} \tag{4.22}$$

as scale-dependent Dirac mollifier scaling functions (see Figure 4.1). Assuming that ρ is suitable, e. g., Hölder-continuous in points of $\mathcal{B}$, then for τ_j-mollifier Newton potentials V_{τ_j} of the gravitational potential V (cf. (3.25)) the limit relation

$$\lim_{j\to\infty} \sup_{x\in\overline{\mathcal{G}}} \left|V(x) - V_{\tau_j}(x)\right| = 0 \tag{4.23}$$

holds for each domain $\mathcal{G}$ with $\mathcal{G} \subset \mathcal{B}$ and $\operatorname{dist}(\partial\mathcal{G}, \partial\mathcal{B}) > 0$.

By application of the negative Laplace operator the transition to the "Haar concept"

$$-\Delta_x G_{\tau_j}(x,y) = H_{\tau_j}(x,y), \tag{4.24}$$

succeeds according to construction whereby the Dirac scaling function H_{τ_j} represents the Haar kernel to the scale j (cf. Figure 4.1)

$$H_{\tau_j}(x,y) = \begin{cases} 0, & |x-y| \ge \tau_j, \\ \frac{3}{4\pi\tau_j^3}, & |x-y| < \tau_j. \end{cases} \tag{4.25}$$

Dirac higher (Taylor) order mollifier scaling functions are given by

$$G^n_{\tau_j}(x,y) = \begin{cases} \frac{n+3}{8\pi\tau_j}, & |x-y| = 0, \\ \frac{1}{4\pi|x-y|} - \frac{(\tau_j - |x-y|)^{n+2}((n+1)|x-y| + 2\tau_j)}{8\pi|x-y|\tau_j^{n+3}}, & 0 < |x-y| < \tau_j, \\ \frac{1}{4\pi|x-y|}, & |x-y| \ge \tau_j, \end{cases} \tag{4.26}$$

such that the n-th order smoothed Haar kernel $H^n_{\tau_j}(x,y)$ defined by

$$H^n_{\tau_j}(x,y) = \begin{cases} 0, & \tau_j \le |x-y|, \\ \frac{(n+1)(n+2)(n+3)}{8\pi} \frac{(\tau_j - |x-y|)^n}{\tau_j^{n+3}}, & 0 \le |x-y| < \tau_j, \end{cases} \tag{4.27}$$

fulfils the differential equation

$$-\Delta_x G^n_{\tau_j}(x,y) = H^n_{\tau_j}(x,y) \tag{4.28}$$

where $G^n_{\tau_j}(x,y)$ is given by (4.26). For Dirac mollifier scaling functions of higher (Taylor) order, the limit relation (4.23) also holds.

To introduce a discrete scale wavelet structure, we consider the sequence of mollifier scaling functions $\{G_{\tau_j}(\cdot,\cdot)\}$, $j \in \mathbb{N}_0$, for the kernel $G(\cdot,\cdot)$ given by (4.22). Corresponding to the family $\{G_{\tau_j}(\cdot,\cdot)\}_{j\in\mathbb{N}_0}$ of mollifier scaling functions, we define the family of mollifier wavelet functions $\{\Psi_{G_{\tau_j}}(\cdot,\cdot)\}_{j\in\mathbb{N}_0}$ using the difference equation (see (4.8))

$$\Psi_{G_{\tau_j}}(x,y) = G_{\tau_{j+1}}(x,y) - G_{\tau_j}(x,y), \quad j \in \mathbb{N}_0, \tag{4.29}$$

$x, y \in \mathbb{R}^3$. Higher order mollified wavelet functions are again defined in a similar way. Written out for the linear case, i. e. $n = 1$, we get

$$\begin{aligned} \Psi_{G_{\tau_j}}(x,y) &= G_{\tau_{j+1}}(x,y) - G_{\tau_j}(x,y) \qquad\qquad (4.30)\\ &= \begin{cases} 0, & |x-y| > \tau_j, \\ \frac{1}{8\pi\tau_j}\left(3 - \frac{|x-y|^2}{\tau_j^2}\right) - \frac{1}{4\pi|x-y|}, & \tau_{j+1} < |x-y| \le \tau_j, \\ \frac{1}{8\pi\tau_{j+1}}\left(3 - \frac{|x-y|^2}{\tau_{j+1}^2}\right) - \frac{1}{8\pi\tau_j}\left(3 - \frac{|x-y|^2}{\tau_j^2}\right), & |x-y| \le \tau_{j+1}. \end{cases} \end{aligned}$$

Again it should be emphasized that each sequence member of $\{\Psi_{G_{\tau_j}}(\cdot,\cdot)\}_{j\in\mathbb{N}_0}$ has a local support, i. e., when evaluating the integral, one can restrict oneself to a certain region, a spherical support with scale-dependent radius τ_j. This makes the calculation efficient and economical from a numerical point of view, as already explained.

Multiscale techniques using mollifier scaling and wavelet functions as constructed in this geomagnetically reflected contribution for explorational application, indeed, represent an outstanding methodology by its particular ability to elaborate specific features of the magnetization distribution by a "zooming-in" process. As an immediate consequence, multiscale mollifier decorrelation may be rated without any doubt as a well-promising and far-reaching methodology in future magnetometric exploration.

However, there is also no doubt that each method in numerics has its own aim and character. In fact, it is the essence of any numerical realization that it becomes optimal only with respect to especially specified criteria. For example, radial basis functions do not show any directional modeling dependence. Trial functions with a local support corresponding to terrestrial-only data systems allow efficient and economical numerics for a near-Earth's surface domain, but if only terrestrial data are available, they do not enable a deeper modeling than the extension radius of the supports, aso. In other words, for the numerical work to be done, we have to make an a priori choice which feature should be pointed out and preferably handled. In particular, for constructive approximation involving mollifier obligations, we have to reflect the different stages of space/frequency localization so that the modeling process can be performed adaptively under the localization requirements necessary and sufficient for geologic interpretation.

Hence, from a numerical point of view, a positive decision in one direction often amounts to a negative implication in another direction. Ultimately, also because of the interdependence of the space/frequency localization caused by the validity of the uncertainty principle in peremptory terms (see, e. g., Freeden2021, [69]), it is impossible to provide a "cure-all methodology". This "sine qua non" ingredient in any constructive mathematical approximation method is also the reason, why we are indispensably led to essential difficulties in the mollifier approximation by scaling and wavelet potentials as proposed here:

- The choice of the type of the mollifier scaling function is an important problem in inverse gravimetry and magnetometry. In this work we restrict ourselves to locally supported trial functions such as Haar-type kernels, but also non-spacelimited trial functions (e. g., Gauss-type functions) are applicable. In principle, all mollifier wavelet strategies are equivalent, however, the suitable computational compromise between mathematical rigor and practical relevance and applicability with respect to the specific geometry of the test area, data width, certain accuracy of the data, and occurring noise level and type is a task, which should not be underestimated.

- In our approach described above, we are not confronted with a multiscale procedure, for which subsequently each detail information guarantees an improvement to come closer and closer to the "real" solution. Well-posed problems usually do not require any stopping strategy of the multiscale (zooming-in) process, since the data are (generally) given in a non-attenuated form. An algorithm establishing an approximate solution for the inverse gravimetry and magnetometry problem, however, has to take into account the requirement to stop at the right level of approximation in order to appropriately model the filtered occurrence of the measured potential data as smooth volume integral values. As a consequence, a missing stopping strategy in the multiscale mollifier regularization usually produces considerable errors in the density or magnetization distribution even from extremely small errors in the potential measurements. In other words, there is a strong need for a relevant and mathematically motivated stopping strategy, which should not be underestimated as well.

4.4 Decorrelation of the BP model

To illustrate the multiscale mollifier technique, we use a realistic model of the gas and oil industry called the BP model. In the 2004 test model (see [25]), the left side is based on a geologic cross-section of the western Gulf of Mexico. The middle panel is a simplified representation of geologic features offshore Angola. The right side is a composite representation of geologic structures encountered in the Caspian Sea, the North Sea, or near Trinidad.

The model is thus a synthetically generated 2D model (see [25]), which for our application was canonically extended into the $\mathbb{R}^3$ in the direction of the x_2-axis to a 3D model (see Figure 4.3). This test model will be used as an example to illustrate decorrelation using the mollifier regularization techniques for a geologic data analysis as a basic requirement for the subsequent inversion.

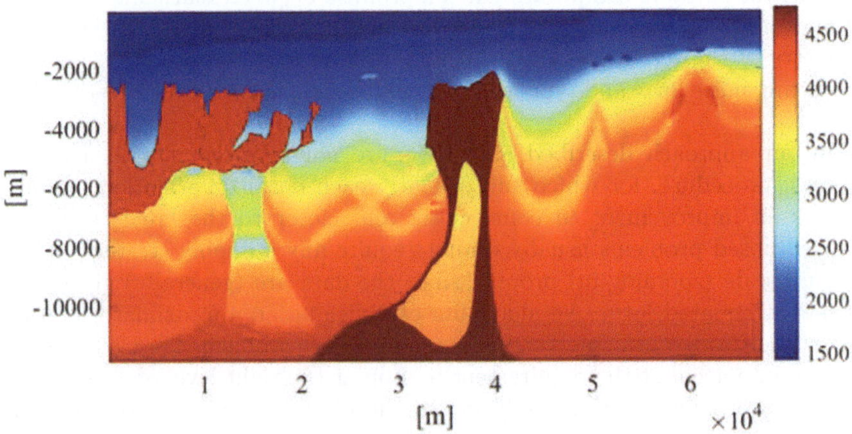

Fig. 4.3 The BP model (adapted from F. J. Billette and S. Brandsberg-Dahl, 2004, The 2004 BP velocity benchmark: 67th Annual EAGE Meeting, EAGE, Expanded Abstracts, Soc. Expl. Geophys., B305).

On the left side in Figure 4.4 the data are reconstructed (lowpass filtering with increasing scales j), on the right side the signatures of the so-called detail components (bandpass filtering with increasing scales j) are seen. The procedure is designed in such a way that the scale j determines the size of the local-compact supports of the wavelets and, thus, is a measure for the filter width of the support wavelets and consequently for the order of magnitude of the bandwidth of the filtered out signatures.

The results convincingly show that as the scale j increases, i. e., as the spatial localization of the wavelets increases, sharper contours of the geologic formation boundaries and discontinuities are revealed. Thus, relevant geothermal information becomes detectable via the decorrelation capability of the mollifier wavelets. The decorrelation procedure using the mollifier functions (4.22), of course, does not provide any knowledge of the type of geologic layers themselves, but rather highly accurate transition surfaces of these geologic layers in their mutual position. Preliminary and/or additional geologic knowledge of the formations separated from each other is therefore indispensable for a complete designation.

Figure 4.5 shows the schematic overview of the decorrelations for the BP gravity and BP density. It is astonishing that already decorrelations of higher scales of the input potential data of V reveal geologic structures without performing the Laplace derivation to determine decorrelations of density signatures ρ. The reason is that difference computation can be interpreted as a discrete Laplace derivative, and thus a discrete version of Poisson's differential equation shows up in the wavelet potential models revealing density signatures.

The use of alternative mollifier functions, e. g. higher order (cf. [62, 63, 64]), may affect the convergence of the multiscale process, but leads in the decorrelation result just as highly accurate to the detection of existing transition surfaces. More detailed investigations as well as comparative tests of data reduction using thresholding can be found (also for alternative density models of the oil and gas industry, e. g., for the Marmousi model) in the monograph "Decorrelative Mollifier Gravimetry – Basics, Concepts, Examples and Perspectives" [54] from 2021.

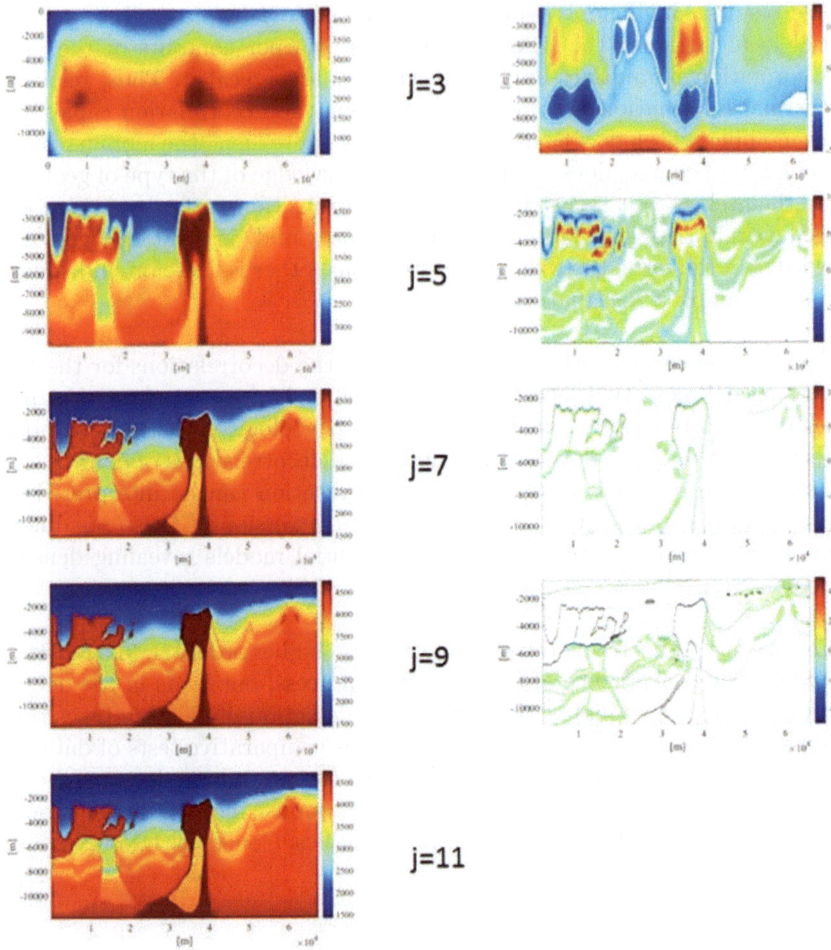

Fig. 4.4 Geologic decorrelation of the BP model from Figure 4.3 based on the inverse problem of gravimetry. In the left column the lowpass components are shown with increasing dyadic scales $j = 3,5,7,9,11$, on the right the bandpass components (detail information) with increasing dyadic scale parameters $j = 3,5,7,9$. The decorrelation procedure provides more and more accurate transition surfaces of the geologic layers in their mutual position with increasing scales.

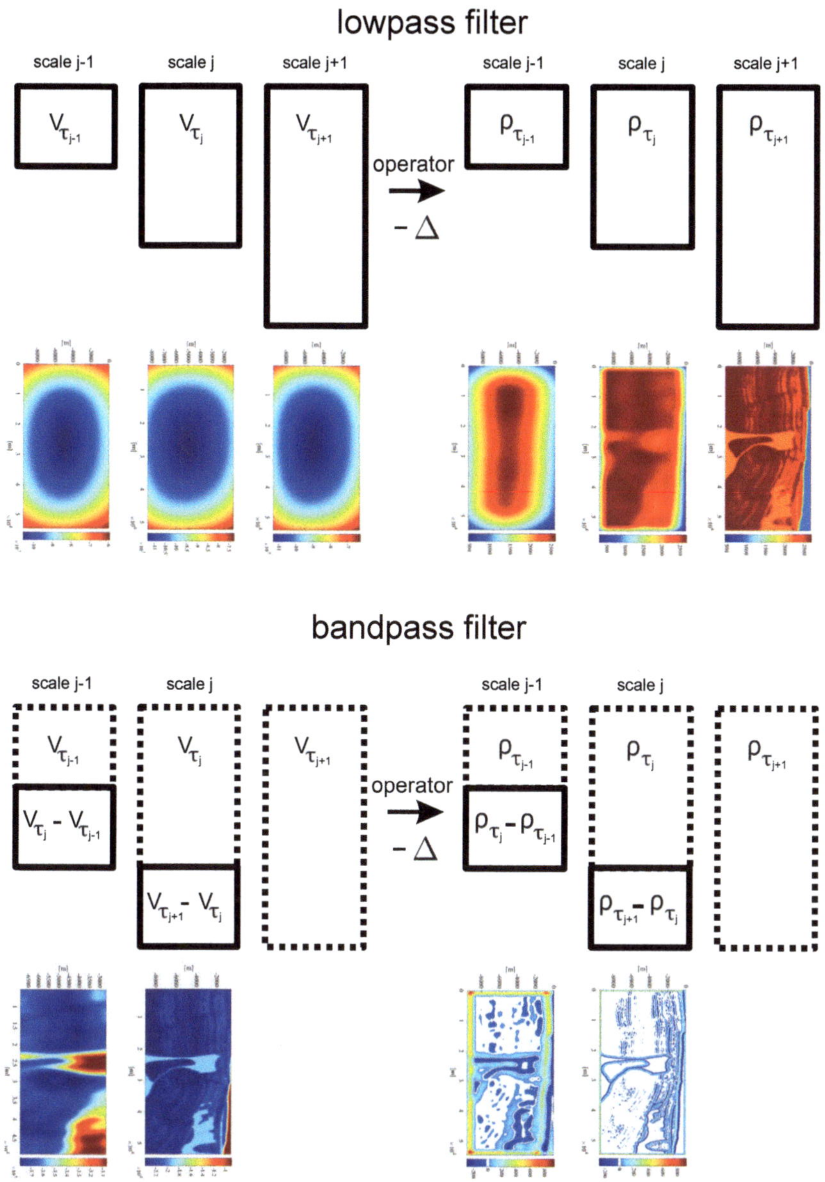

Fig. 4.5 Geologic decorrelations of the gravitational potential and density of the BP model from Figure 4.3 based on gravimetric wavelet decomposition. It is remarkable that the density structure can be seen also in subsequent differences of the gravitational potential, since the discretization of the Laplace operator leads to an associated difference operator, hence, to a discrete version of Poisson's equation.

Chapter 5

Magnetometric multiscale approximation and mollifier decorrelation

Abstract:

The key idea of the mollifier multiscale method is that the corresponding potential wavelets generating the bandpass decomposition of the potential can be simultaneously transformed via the application of a differential operator to wavelet decompositions in the contrast function level. Remarkably, this chapter shows that this can be implemented for both density and magnetization structures in respective parallelism. In fact, our multiscale mollifier methodology enables us to represent the contrast function, but especially its decorrelation into distribution bands, in such a way that sharp contrast transitions from one geologic formation to another can be detected better and better with increasing scales.

5.1 Concept of magnetometric mollifier regularization

By letting the distance of two monopoles converge towards zero, a difference quotient is mathematically transformed into a differential quotient and by integration over the test area $\mathcal{B}$ we finally obtain

$$B(x) = \frac{1}{4\pi} \int_{\mathcal{B}} \nabla_y \frac{1}{|x-y|} \cdot m(y) \, dV(y). \tag{5.1}$$

W. Freeden, H. Nutz, *Exploratory Potential Methods in Geothermal Power Generation*, Lecture Notes in Geosystems Mathematics and Computing,
https://doi.org/10.1007/978-3-031-54412-5_5

So, as a result, the magnetic potential B of the dipole is representable as the infinitesimal sum of two monopole potentials (cf., e. g., [13, 14, 15, 27, 34, 155, 156]).

The integral (5.1) can also be expressed as follows:

$$\begin{aligned} B(x) &= -\frac{1}{4\pi}\int_{\mathcal{B}} \frac{m(y)\cdot(x-y)}{|x-y|^3}\, dV(y) \\ &= \frac{1}{4\pi}\int_{\mathcal{B}} \nabla_x\cdot\left(-\frac{1}{|x-y|}\, m(y)\right) dV(y) \\ &= -\nabla_x\cdot\frac{1}{4\pi}\int_{\mathcal{B}} m(y)\,\frac{1}{|x-y|}\, dV(y). \end{aligned}$$

The magnetic induction b is the gradient field of the dipole potential B of a magnetization m, i. e., b is given by $b=\nabla B$ with

$$b(x) = -\nabla_x \frac{1}{4\pi}\int_{\mathcal{B}} \frac{m(y)\cdot(x-y)}{|x-y|^3}\, dV(y), \quad x\in\mathbb{R}^3. \tag{5.2}$$

For points $x\in\mathbb{R}^3\backslash\overline{\mathcal{B}}$

$$\begin{aligned} b(x) &= -\frac{1}{4\pi}\int_{\mathcal{B}} \nabla_x\frac{m(y)\cdot(x-y)}{|x-y|^3}\, dV(y) \\ &= -\frac{1}{4\pi}\int_{\mathcal{B}}\left(\frac{m(y)}{|x-y|^3} - 3\frac{(x-y)\cdot m(y)}{|x-y|^5}(x-y)\right) dV(y) \end{aligned} \tag{5.3}$$

is valid.

An elementary calculation shows that the magnetic field b is divergence-free outside $\mathcal{B}$, i. e., for points $x\in\mathbb{R}^3\backslash\overline{\mathcal{B}}$

$$\nabla\cdot b(x) = \nabla\cdot\nabla B(x) = \Delta B(x) = 0, \quad x\in\mathbb{R}^3\backslash\overline{\mathcal{B}}, \tag{5.4}$$

holds. For points $x\in\mathbb{R}^3\backslash\overline{\mathcal{B}}$

$$\int_{\mathcal{B}} m(y)\cdot\nabla_y\frac{1}{|x-y|}\, dV(y) = \int_{\mathcal{B}}\left(\nabla_y\cdot\frac{m(y)}{|x-y|} - \frac{1}{|x-y|}\nabla_y\cdot m(y)\right) dV(y) \tag{5.5}$$

is still valid.

In connection with the classical theorem of Gauss of vector analysis we obtain the following result: Let m be continuously differentiable over $\overline{\mathcal{B}}$. Then, for $x\in\mathbb{R}^3\backslash\overline{\mathcal{B}}$, we can obtain the potential

$$B(x) = \frac{1}{4\pi}\int_{\mathcal{B}} m(y)\cdot\nabla_y\frac{1}{|x-y|}\, dV(y) \tag{5.6}$$

express by a volume and surface integral in the form

$$\begin{aligned} B(x) = & -\frac{1}{4\pi}\int_{\mathcal{B}}\frac{1}{|x-y|}\nabla_y\cdot m(y)\, dV(y) \\ & +\frac{1}{4\pi}\int_{\partial\mathcal{B}}\frac{1}{|x-y|}\, m(y)\cdot\nu(y)\, dS(y), \end{aligned} \tag{5.7}$$

where ν is the unit normal field pointing to the exterior of $\overline{\mathcal{B}}$.

The representation (5.7) leads us to the consequence that the magnetic induction $b = \nabla B$ in free space outside a magnetized body $\overline{\mathcal{B}}$ provides information about the divergence of the magnetization and not the magnetization itself. Therefore, any magnetization distribution that is divergence-free is magnetically quiescent, so it cannot affect the external magnetic field. Moreover, the decomposition (5.7) into a surface and a volume integral shows that the contribution to the potential B in $\overline{\mathcal{B}}$ comes only from sources and sinks of magnetization in $\mathcal{B}$ with $\nabla\cdot m \neq 0$, and from parts of the surface $\partial\mathcal{B}$, where the surface normals ν and m are not orthogonal.

5.2 Magnetic field and susceptibility

Next, we follow the standard literature in geomagnetics (cf., e. g., [27, 124]) to characterize the macroscopic framework for our planned modeling: The principle to be realized is that at distances large compared to the spatial extent, any current distribution produces a magnetic induction that can be approximately understood as a magnetic dipole. Thus, we can formulate the basic equation for b as follows (see, e. g., [34]):

$$\nabla \underbrace{(b-m)}_{=h} = j, \tag{5.8}$$

where j is the current density and a new field h, called the *macroscopic magnetic field*, is given by

$$h = b - m. \tag{5.9}$$

Accordingly, macroscopic magnetostatics is characterized by a relation between b and h which, according to the formula (5.9), corresponds to a relation between m and h (or b). In fact, for a large number of materials there is an approximately linear relationship between m and h. When an object of magnetically linear and additionally isotropic material is placed in a magnetic field generated by a distant electromagnet, the magnetization $m(x)$ at a source point x is given by the product of the *magnetic susceptibility* $\chi(x)$ and the magnetic field $h(x)$,, i. e.,

$$m(x) = \chi(x)\, h(x). \tag{5.10}$$

The magnetic susceptibility is called *paramagnetic* when $\chi > 0$; the magnetic susceptibility is called *diamagnetic* when $\chi < 0$. Normally, χ is quite small for paramagnetic and diamagnetic materials (i. e., $|\chi(x)| \simeq 10^{-5}$).

A linear relation between h and m implies also a linear relation between b and h. In fact, the magnetic dipole induction field b at the same source point x is given by

$$b(x) = h(x) + m(x). \tag{5.11}$$

It should not go unmentioned that the *permeability of free space* μ_0 is set equal to one here, unlike in numerical implementations. Consequently, we can express (5.10) in terms of the susceptibility χ by substituting (5.10) into (5.11) so that

$$b(x) = (1+\chi(x))\, h(x) = \mu(x)\, h(x), \tag{5.12}$$

where the (absolute) *permeability* μ is given by $\mu(x) = 1+\chi(x)$ (for more details compare also [34]).

5.3 Born approximation for the magnetometric integral equation

We know that the magnetization of a body $\overline{\mathcal{B}}$ is related to the magnetic field h at location x as follows: $m(x) = \chi(x)\, h(x)$. Assuming that the magnetization of the body is parallel to the inducing field (cf. [34] for more details), we can assume magnitude and direction proportional to the susceptibility χ. This leads to the so-called *Born approximation for the magnetometric integral equation*

$$b(x) = \nabla_x \int_{\mathcal{B}} \underbrace{\nabla_y G(x,y)}_{=\gamma(x,y)} \cdot \underbrace{m(y)}_{=\chi(y)h(y)}\; dV(y). \tag{5.13}$$

The goal now is to calculate a possible distribution of the susceptibility χ below the surface that could have led to the anomalous fields in the magnetic survey data. Again, this is an inverse problem of ill-posed nature to which the mollifier methodology is applicable in a well-known way (comparable to the gravimetric case). In geophysical practice, one often works with input data in a certain direction l, i. e., instead of b the scalar product $l \cdot b$ is used. If the data is given in free space (i. e., air), then $\mu \simeq 1$. So, we treat $l \cdot h$ as if it were $l \cdot b$. In conjunction with (5.12), we are then faced with the inverse problem of determining the susceptibility χ from scalar data of $l \cdot b$:

$$l(x) \cdot b(x) \simeq (l(x) \cdot \nabla_x) \int_{\mathcal{B}} (\gamma(x,y) \cdot b(y))\chi(y)\, dV(y). \tag{5.14}$$

Obviously, the "inverse magnetometric Born problem" (5.14) allows a similar mollifier methodology as scalar inverse gravimetry (see [54]).

5.4 Magnetometric mollifier functions

Instead of determining the density signatures from the gravitational potential in gravimetry, the magnetometric case is characterized by seeking the determination of the vector magnetization m from the magnetic field potential B using the scalar Fredholm integral equation

$$B(x) = \int_{\mathcal{B}} \underbrace{m(y) \cdot \nabla_y \underbrace{\frac{1}{4\pi|x-y|}}_{=G(x,y)}}_{=\gamma(x,y)}\, dV(y). \tag{5.15}$$

As already mentioned, the magnetometric integral equation connects the dipole potential B with the magnetization m via a vectorial integral kernel $\gamma(x,y)$, which arises from $G(x,y)$ by calculation of the gradient

$$\gamma(x,y) = \nabla_y G(x,y) = -\frac{x-y}{4\pi|x-y|^3} = -\nabla_x G(x,y), \quad x \neq y. \tag{5.16}$$

Formally, the integrand of (5.15) here is to be understood as the product between the sought magnetization m and the integral kernel $\gamma(x,y)$, i. e., as an inner product between two vector valued functions.

Decorrelation formula for the dipole potential. For a monotonically decreasing zero sequence $\{\tau_j\}_{j\in\mathbb{N}_0}$ the "mollification" of the integral kernel $\gamma(x,y)$, including higher order mollifiers of the $(n+1)$-fold continuously differentiable kernel $G^n_{\tau_j}(x,y)$ takes the form $\gamma^n_{\tau_j}(x,y) = \nabla_y G^n_{\tau_j}(x,y)$ according to (4.26), i. e., we have

$$\gamma^n_{\tau_j}(x,y) = \begin{cases} -\frac{x-y}{4\pi|x-y|^3}, & \tau_j \le |x-y|, \\ -\frac{x-y}{4\pi|x-y|^3} - \frac{(x-y)(\tau_j-|x-y|)^{n+2}(n+1)}{8\pi|x-y|^2{\tau_j}^{n+3}} \\ \quad +\frac{(x-y)(\tau_j-|x-y|)^{n+2}(2\tau_j+|x-y|(n+1))}{8\pi|x-y|^3{\tau_j}^{n+3}} \\ \quad +\frac{(x-y)(\tau_j-|x-y|)^{n+1}(n+2)(8\pi\tau_j+|x-y|(n+1))}{8\pi|x-y|^2{\tau_j}^{n+3}}, & 0 < |x-y| < \tau_j, \\ 0, & |x-y| = 0, \end{cases} \tag{5.17}$$

(see Figure 5.1 for profile representations).

The developed mollifiers $\gamma^n_{\tau_j}(x,y)$ now canonically lead to a multiscale decomposition. The associated wavelets are then obtained as the difference of two mollifiers at scales $j+1$ and j (cf. Figure 5.2). At the same time, this means that a mollifier to scale J can be represented as the sum of a mollifier lowpass filter to scale $j_0 < J$ and the mollifier wavelets to scales $j = j_0, \ldots, J-1$. The mathematical convolution of the magnetization with this sum now provides the multiscale decomposition of the magnetic field potential. In summary, as in the case of gravimetry, convolution with the mollifier to scale j_0 yields lowpass filtering and convolution with the wavelets yields corresponding bandpass filtering. This filtering is used with increasing level of detail for the geologic interpretation and, at the same time, it provides the basis for the desired inversion process.

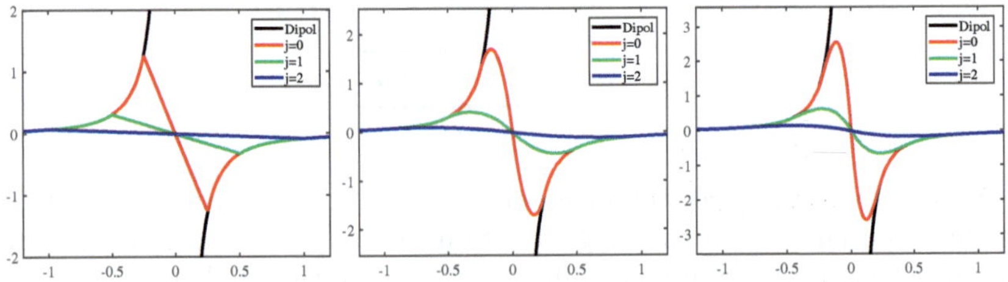

Fig. 5.1 Profile line of the first component of the vector-valued integral kernel $\gamma(x,0)$ (black) and its regularization $\gamma^n_{\tau_j}(x,0)$ for $\tau_j = 1$, 0.5, 0.25, $n=0$ (left), $n=1$ (middle) and $n=2$ (right) in $x_2, x_3 = 0$.

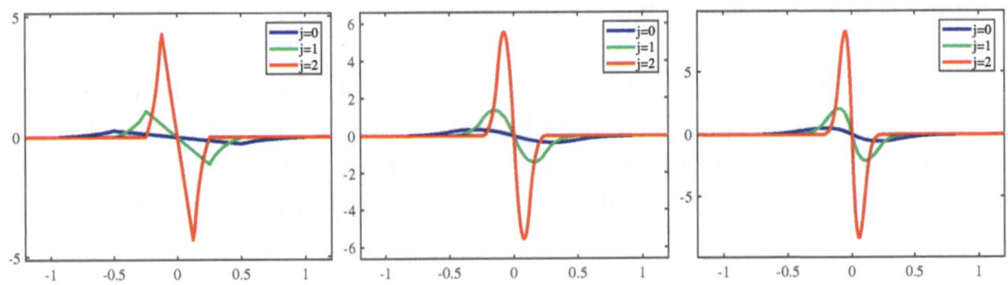

Fig. 5.2 Profile line of the first component of wavelets $\gamma^n_{\tau_{j+1}}(x,0) - \gamma^n_{\tau_j}(x,0)$ for $\tau_j = 2^{-j}$ and $n=0$ (left), $n=1$ (middle) and $n=2$ (right) in $x_2, x_3 = 0$.

In mathematical formula representation we thus obtain a lowpass filtering of the dipole potential in the mollifier form

$$B^n_{\tau_j}(x) = \int_{\mathcal{B}} \gamma^n_{\tau_j}(x,y) \cdot m(y)\, dV(y). \tag{5.18}$$

By subtracting two lowpass filtered dipole potentials to scales j and $j-1$, i. e., $(WB)^n_{\tau_j} = B^n_{\tau_j} - B^n_{\tau_{j-1}}$ we obtain bandpass filtered dipole potentials $(WB)^n_{\tau_j}$, where we may assume approximately that $B(x) \approx (B)^n_{\tau_J}(x)$ for a suitably high scale $J > j_0$. This leads to the following approximate wavelet reconstruction formula

$$B(x) \approx B^n_{\tau_J}(x) = B^n_{\tau_0}(x) + \sum_{j=0}^{J} (WB)^n_{\tau_j}(x), \tag{5.19}$$

from which a decorrelation of $B \approx B^n_{\tau_J}$ is obtained by studying $B^n_{\tau_0}$ and $(WB)^n_{\tau_j}$. According to the multiscale access (5.19), the dipole potential B is thus composed of a lowpass filtering $B^n_{\tau_0}$ plus detail information in the form of bandpass filtered dipole potentials $(WB)^n_{\tau_j}(x)$, $j = 0, \ldots, J$, at each location $x \in \mathcal{B}$.

Decorrelation formula for dipole induction. For the development of mollifers for magnetic dipole induction b, we apply the gradient to the mollification of the dipole potential, i. e., $b^n_{\tau_j}(x) = (\nabla B)^n_{\tau_j}(x)$. Thus we obtain a lowpass filtered mollifier dipole induction $b^n_{\tau_j}(x)$ by means of

$$b^n_{\tau_j}(x) = \int_{\mathcal{B}} \mathfrak{g}^n_{\tau_j}(x,y)\; m(y)\, dV(y). \tag{5.20}$$

Here the tensor scaling function $\mathfrak{g}^n_{\tau_j}(x,y)$ is given by

$$\mathfrak{g}^n_{\tau_j}(x,y) = \nabla_x \gamma^n_{\tau_j}(x,y) \tag{5.21}$$

$$= \begin{cases} (x-y)(x-y)^T \frac{3}{4|x-y|^5\pi} - \mathbf{i}\, \frac{1}{4\pi|x-y|^3}, & \tau_{j_j} \le |x-y|, \\ -(x-y)(x-y)^T \Bigg(\frac{6\tau_j{}^3(\tau_j-|x-y|)^n - 6\tau_j{}^{n+3}}{8|x-y|^5\tau_j{}^{n+3}\pi} \\ \quad + \frac{3|x-y|^3 n^2(\tau_j-|x-y|)^n}{8|x-y|^5\tau_j{}^{n+3}\pi} \\ \quad + \frac{|x-y|^3 n^3(\tau_j-|x-y|)^n + 2|x-y|^3 n(\tau_j-|x-y|)^n}{8|x-y|^5\tau_j{}^{n+3}\pi} \\ \quad + \frac{6|x-y|n\tau_j{}^2(\tau_j-|x-y|)^n + 3|x-y|^2 n\tau_j(\tau_j-|x-y|)^n}{8|x-y|^5\tau_j{}^{n+3}\pi} \\ \quad + \frac{3|x-y|^2 n^2\tau_j(\tau_j-|x-y|)^n}{8|x-y|^5\tau_j{}^{n+3}\pi} \Bigg) \\ +\mathbf{i} \Bigg(\frac{(\tau_j-|x-y|)^{n+2}(|x-y|+2\tau_j+|x-y|n)}{8\pi|x-y|^3\tau_j n+3} - \frac{1}{4\pi|x-y|^3} \\ \quad - \frac{(\tau_j-|x-y|)^{n+2}(n+1)}{8\pi|x-y|^2\tau_j{}^{n+3}} \\ \quad + \frac{(\tau_j-|x-y|)^{n+1}(n+2)(|x-y|+2\tau_j+|x-y|n)}{8\pi|x-y|^2\tau_j{}^{n+3}} \Bigg), & 0 < |x-y| < \tau_j, \\ -\mathbf{i}\, \frac{n^3+6n^2+11n+6}{24\pi\tau_j{}^3}, & 0 = |x-y|, \end{cases}$$

where $\mathbf{i}$ denotes the unit matrix.

The bandpass filtering is again generated by difference of lowpass filterings of successive scales $(Wb)^n_{\tau_j} = b^n_{\tau_j} - b^n_{\tau_{j-1}}$. For each $J > 0$ we thus obtain a reconstruction of $b^n_{\tau_J}$ by means of

$$b^n_{\tau_J}(x) = b^n_{\tau_0}(x) + \sum_{j=0}^{J} (Wb)^n_{\tau_j}(x). \tag{5.22}$$

Decorrelation formula for the divergence of magnetization. The decorrelation formula for the divergence of magnetization $M = \nabla \cdot m = \operatorname{div} m$ is used to show magnetic sources and sinks in a dataset m. These then – as will be shown – also provide information about geologic specifics.

We already know that M can be calculated from the dipole potential by the negative Laplace derivative, so that

$$M^n_{\tau_j}(x) = -\Delta B^n_{\tau_j}(x) = -\nabla_x \cdot b^n_{\tau_j}(x) = -\int_{\mathcal{B}} H^n_{\tau_j}(|x-y|)\nabla_y \cdot m(y)\, dV(y). \tag{5.23}$$

Again, we obtain a bandpass filtering by $(WM)^n_{\tau_j} = M^n_{\tau_j} - M^n_{\tau_{j-1}}$, which leads to the reconstruction formula

$$M^n_{\tau_J}(x) = M^n_{\tau_0}(x) + \sum_{j=0}^{J} (WM)^n_{\tau_j}(x) \tag{5.24}$$

leading to the determination of the decorrelation levels $(WM)^n_{\tau_j}$, $j = 0, \ldots, J$.

5.5 Decorrelation of the BP model

As in the case of gravimetry, the decorrelation will now be illustrated in more detail using the example of the BP model (see Figure 5.3). The magnetic field and the divergence of the magnetization are shown in a multiscale scheme.

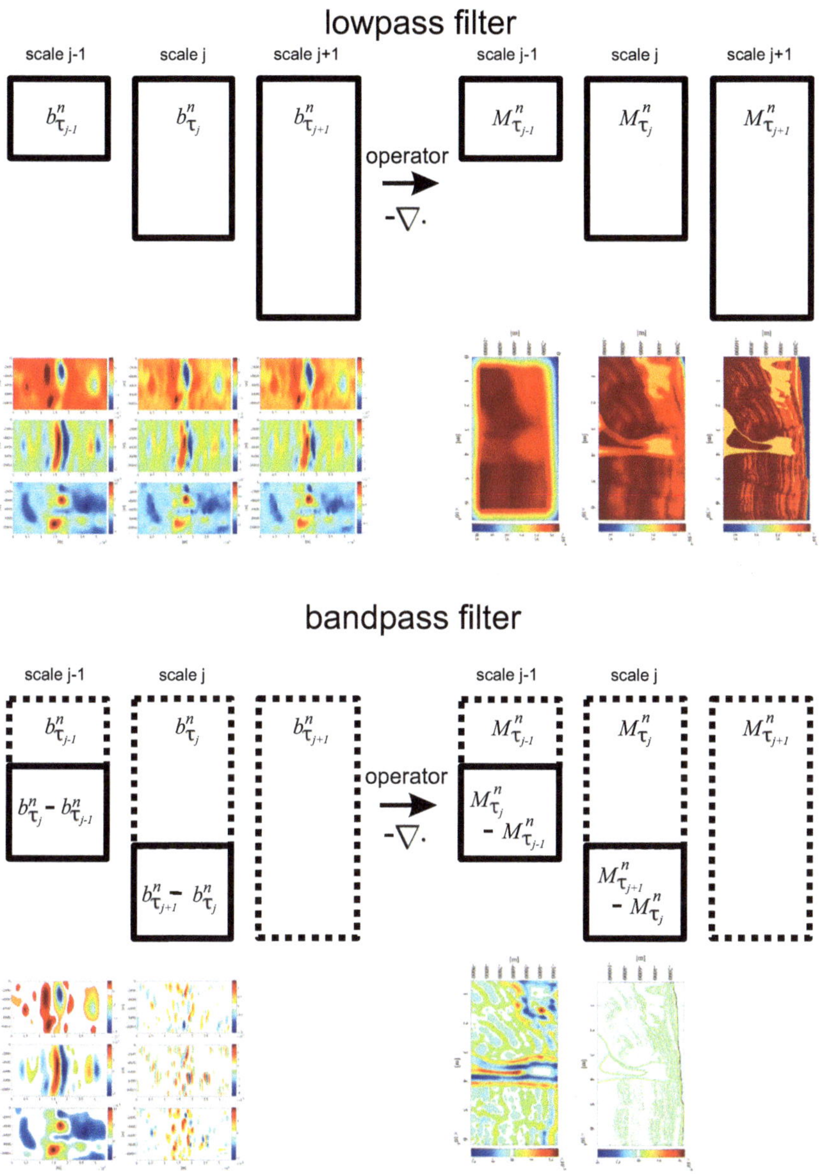

Fig. 5.3 Geologic decorrelation of the magnetic field (left) of the BP model from figure 4.3 and the divergence of magnetization (right) based on magnetic wavelet decomposition.

Chapter 6

Test reservoir modeling in Bavarian molasse basin

Abstract:

This chapter takes the ideas and concepts developed in the books [34, 54] and briefly recapitulated earlier in this work about gravimetric and/or magnetometric multiscale mollifier approximation. The approaches are realized numerically in an (application-oriented) wavelet as well as spline wavelet model. As a geologically relevant test example, a real dataset from the Bavarian Mollasse basin from 2012 is used, which was generated in the course of the construction of the geothermal power plant Traunreut.

6.1 Test area Traunreut in the Bavarian molasse basin

The *Molasse basin* (see also [54]) is a foreland basin north of the Alps which was formed during the Oligocene and Miocene epochs. The basin represents the result of the flexure of the European plate under the weight of the orogenic wedge of the Alps that was forming to the south.

The Bavarian Molasse Basin has a special significance in geothermal energy. In the central and eastern part of the Molasse Basin, the Malm (Upper Jurassic) is the most potentially productive thermal water aquifer. Under particularly favorable circumstances, energy utilization in the Cretaceous sandstones could also be possible. Following [122], the Malm reaches the greatest thicknesses under the Molasse Basin with more than 600 m south

W. Freeden, H. Nutz, *Exploratory Potential Methods in Geothermal Power Generation*, Lecture Notes in Geosystems Mathematics and Computing,
https://doi.org/10.1007/978-3-031-54412-5_6

of Munich between Lech and Inn. It is composed up to 500 m thick, massive sponge as well as light, porous coral rubble limestones deposited above them. On the northern slope, uplifted by the Bohemian Massif, several hundred meters of Malm sediments were eroded, so that today, in the eastern part of the Bavarian Molasse Basin, the thicknesses towards the Northeast decrease to less than 100 m. The movement of groundwater in the Malm is essentially along karst cavities, fractures and fault zones and, to a lesser extent, along bed joints. The facial structure of the Malm has a different influence on the groundwater flow due to the different karstability, the facies structure of the Malm has a direct influence on its permeability. As a rule, the dolomitized mass limestones are well permeable, because recrystallization led to an increase in porosity and, in addition, karstification at the grain boundaries (due to [2]).

As a consequence, most of Germany's deep geothermal systems are currently in the Malm aquifers of the Bavarian Molasse Basin. The Malm layer falls from the Danube, where it is at the surface of the Earth, towards the Alps. The geothermal relevance is caused by the fact that the Malm aquifer is filled with water due to its porosity/permeability. Locally, the Malm can be either banky or reef-like (facies) and then either fissured and/or karstified. All of these phenomena can overlap and are the subject of detailed exploration within the last decades. Because of the increasing depth towards the south, the water temperature usually increases. As a result the Malm north of Munich is only used in the heating market, south of Munich also for generating electricity. As a matter of fact, the water of the Malm generally is of drinking water quality and the salt content is very low (more details about the Bavarian Molasse Basin can be found, e. g., in [21, 22, 23, 26, 39, 125, 165] and the references therein).

We test the proposed mollifier decorrelation method on a particular dataset available as seismic records for a certain area of the Bavarian Molasse Basin in the southern Germany. More explicitly, for test calculations with the mollifier inversion methods, a density dataset from preliminary seismic investigations of the Traunreut geothermal project in the Molasse Basin in southern Germany is used. Furthermore, for magnetometric purposes, we consider the known 3D-density distribution signature of the Bavarian Molasse area as the divergence of the magnetization distribution. In fact, this (artificial) setup provides a realistic test dataset varying in all the spatial directions, while guaranteeing a concrete association to geologic exploration. The dimension of the dataset amounts to $92 \times 60 \times 58$ data points situated on a regular grid of 50 m in each direction. The vectorial test magnetization m is obtained from the scalar "data input" by discretizing the divergence operator by a finite difference scheme leading to a linear equation system to be solved for the magnetization from its corresponding divergence.

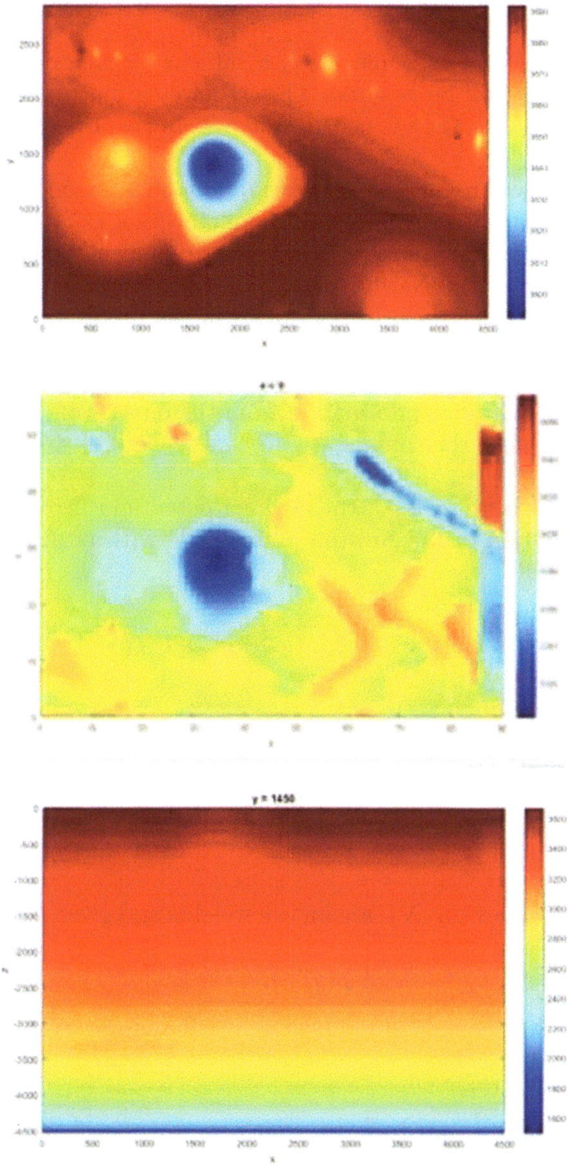

Fig. 6.1 Representation of the density of the test dataset used for the Traunreut geothermal project (top: Density profile below the surface, middle: Density profile at a depth of about 400 m, bottom: Cross section of the density profile in east-west direction).

6.2 Gravimetric wavelet mollifier methodology

For a representation of the subsurface, a sufficiently high density and accuracy of the data must be ensured. This requires a much tighter and more accurate measurement than is usually the case in geophysics and geodesy.

In the sequel, we are interested in synoptic mollifier modeling in gravimetry and magnetometry. We start with the methodology of gravimetric mollifier inversion as developed in [54]. The numerical calculation of the mollifier volume potentials of the Newton integral is based on essential tools of approximate integration, which can be found in the literature, e. g., [58]. The mollifier approach used leads to regularizations which counteract the solution problem of ill-posedness (particularly, non-uniqueness) under knowledge of certain additional prior information from the interior of the area of investigation, e. g., from earlier mining.

For sufficiently large scales J, the mollifier potential $V^n_{\tau_J}$ can be understood as an approximation of the Newton potential V with negligible error, i. e., we have (cf. (4.5))

$$V(x) \approx V^n_{\tau_J}(x) = \int_{\mathcal{B}} G^n_{\tau_J}(x,y)\ \rho(y)\ dV(y), \quad x \in \overline{\mathcal{B}}. \tag{6.1}$$

A cubature formula over $\overline{\mathcal{B}}$ then yields

$$V(x) \approx \sum_{i=1}^{N_J} G^n_{\tau_J}(x, y_i^{N_J})\ w_i^{N_J}\ \rho(y_i^{N_J}), \tag{6.2}$$

where $w_i^{N_J}, y_i^{N_J}, i = 1, \ldots, N_J$, are known weights and given grid points. Since the coefficients

$$a_i^{N_J} = w_i^{N_J}\ \rho(y_i^{N_J}), \quad i = 1, \ldots, N_J, \tag{6.3}$$

are unknown, they must be obtained from the known gravitational values $V(x_k^{M_J})$ at the locations $x_k^{M_J} \in \overline{\mathcal{B}}$, $k = 1, \ldots, M_J$, by solving a linear system of equations of the type

$$V(x_k^{M_J}) = \sum_{i=1}^{N_J} G^n_{\tau_J}(x_k^{M_J}, y_i^{N_J})\ a_i^{N_J}, \quad k = 1, \ldots, M_J. \tag{6.4}$$

Since the integration weights $w_i^{N_J}$ are known, the density values $\rho(y_i^{N_J})$ are immediately obtained from (6.3), so that the density ρ can be determined from

$$\rho(x) \simeq \rho^n_{\tau_J}(x) = \sum_{i=1}^{N_J} H^n_{\tau_J}(x, y_i^{N_J})\, w_i^{N_J} \rho(y_i^{N_J}), \quad x \in \mathcal{B}. \tag{6.5}$$

With the solution of the system of equations, the solution of the gravimetric problem of determining decorrelative density information from discrete potential data is thus also guaranteed at the same time:

$$\rho^n_{\tau_J}(x) = \sum_{i=1}^{N_J} \underbrace{(-\Delta G^n_{\tau_J}(x, y_i^{N_J}))}_{=H^n_{\tau_J}(x, y_i^{N_J})}\, a_i^{N_J} \tag{6.6}$$

and

$$(W\rho)^n_{\tau_J}(x) = \sum_{i=1}^{N_J} \underbrace{(-\Delta (WG)^n_{\tau_J}(x, y_i^{N_J})}_{=(WH)^n_{\tau_J}(x, y_i^{N_J})}\, a_i^{N_J}. \tag{6.7}$$

This allows a mollifier decorrelation analysis of the density ρ from given potential data.

Inclusion of prior information. To address the difficulties of non-uniqueness of the solution and to avoid the fully occupied coefficient matrix of the linear system (6.4), we can assume that from terrestrial (in combination with available internal) prior information, the potential V can be assigned to a known lowpass filtered potential $V^n_{\tau_{J_0}}$, $J_0 < J$.

We consider a potential with a smoothed Newton kernel

$$V^n_{\tau_{J_0}}(x) \;=\; \int_{\mathcal{B}} G^n_{\tau_{J_0}}(x, y)\, \rho(y)\, dV(y). \tag{6.8}$$

In this case, for a sufficiently large J and for all $x \in \mathbb{R}^3$, it holds that

$$\begin{aligned} V(x) - V^n_{\tau_{J_0}}(x) &\simeq V^n_{\tau_J}(x) - V^n_{\tau_{J_0}}(x) \\ &= \int_{\mathcal{B}} \left(G^n_{\tau_J}(x, y) - G^n_{\tau_{J_0}}(x, y) \right) \rho(y)\, dV(y). \end{aligned} \tag{6.9}$$

The difference kernel appearing in the integral of (6.9) has a local support. More precisely, for all $x \in \mathbb{R}^3$ with $|x - y| \geq \tau_{J_0}$, we have

$$G^n_{\tau_J}(x, y) - G^n_{\tau_{J_0}}(x, y) \;=\; 0. \tag{6.10}$$

This property is of great importance for the numerical implementation.

With a suitable cubature formula over $\overline{\mathcal{B}}$ we get

$$V(x) - V^n_{\tau_{J_0}}(x) \simeq \sum_{i=1}^{N_J} \left(G^n_{\tau_J}(x, y_i^{N_J}) - G^n_{\tau_{J_0}}(x, y_i^{N_J}) \right) w_i^{N_J} \; \rho(y_i^{N_J}),$$

where $w_i^{N_J} \in \mathbb{R}$, $y_i^{N_J} \in \mathbb{B}_{\tau_{J_0}}(x)$, $i = 1, \ldots, N_J$, are the known weights and nodes, respectively. The coefficients

$$a_i^{N_J} = w_i^{N_J} \; \rho(y_i^{N_J}), \quad i = 1, \ldots, N_J, \tag{6.11}$$

are then calculated using the known values

$$V(x_k^{M_J}) - V^n_{\tau_{J_0}}(x_k^{M_J}), \quad x_k^{M_J} \in \mathbb{R}^3, \quad k = 1, \ldots, M_J, \tag{6.12}$$

from the linear equation system

$$V(x_k^{M_J}) - V^n_{\tau_{J_0}}(x_k^{M_J}) = \sum_{i=1}^{N_J} \left(G^n_{\tau_J}(x_k^{M_J}, y_i^{N_J}) - G^n_{\tau_{J_0}}(x_k^{M_J}, y_i^{N_J}) \right) a_i^{N_J}, \tag{6.13}$$

$k = 1, \ldots, M_J$, which numerically allows the property (6.10).

Even more (cf. [54]), the efficiency and economy of the linear system (6.13) depends essentially on the distribution and density of data points $x_k^{M_J}$, $k = 1, \ldots, M_J$, for the available potential values $V(x_k^{M_J})$, $k = 1, \ldots, M_J$:

- If the potential values for data points $x_k^{M_J}$, $k = 1, \ldots, M_J$, are available in "appropriately distributed" density over $\overline{\mathcal{B}}$, the more prior is information available, i. e., the larger J_0 with $J_0 < J$ can be chosen, the higher the efficiency.
- For exclusively given terrestrial potential values (i. e., $x_k^{M_J}$, $k = 1, \ldots, M_J$, are points of the intersection of the Earth's surface with the boundary $\partial\mathcal{B}$), however, the local support of the mollifier wavelet function implies a limitation, so that we cannot perform deeper geologic modeling due to the limited extent of the support. If one is interested in exploring greater depths, the support must be adjusted, i. e., J_0 must be chosen appropriately small, which of course affects the numerical economy.

As a consequence, we are confronted with an unfortunate situation, typical for inverse problems, that numerical effectiveness and modeling in greater depth are mutually exclusive and inevitably require a compromise.

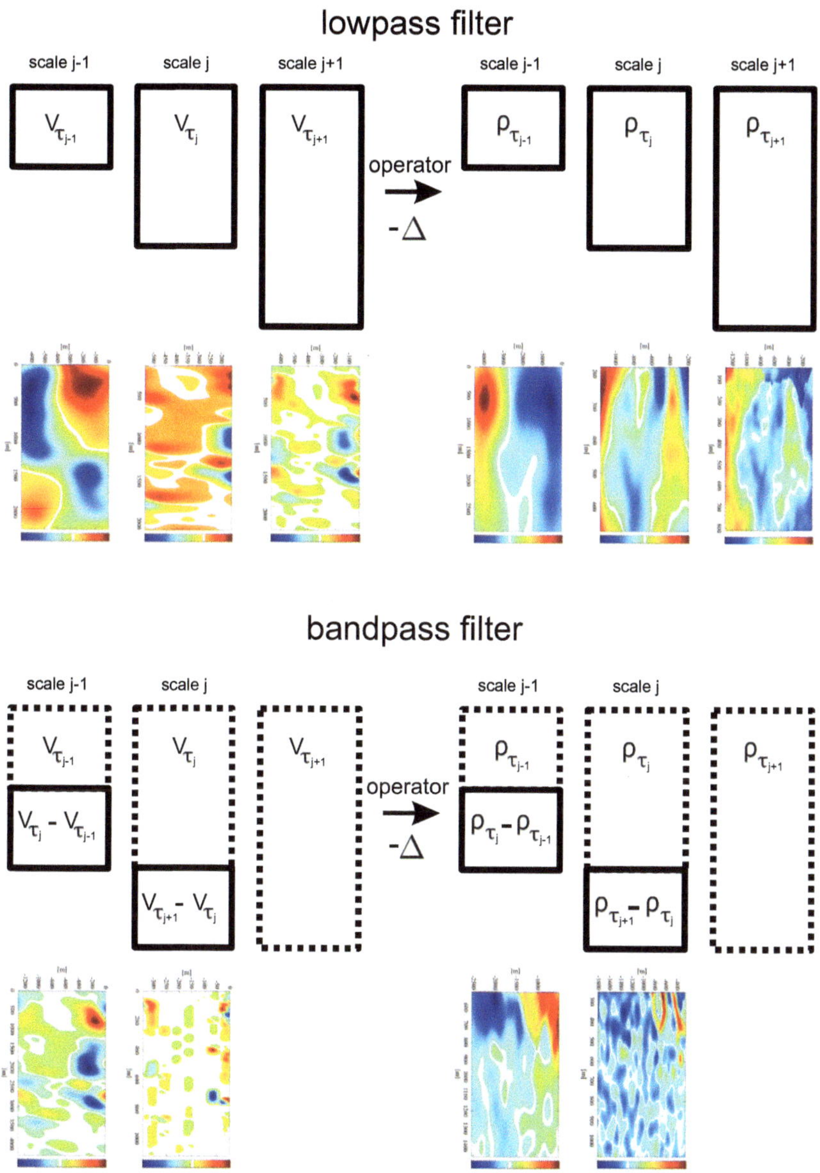

Fig. 6.2 Schematic correlation of decorrelations of gravitational potential and density using the example of Traunreut, Bavarian Molasse Basin.

Figure 6.2 shows the schematic overview of the decorrelations of gravitational potential and density using the example of the Traunreut geothermal project.

6.3 Magnetometric wavelet mollifier methodology

There are several ways to obtain approximations for magnetic induction, of which we examine only basic approaches in more detail here (for more information, see [34, 54]):

Tensor-based variant. Due to $\mathfrak{b} = \nabla B$, we are led canonically to the equation

$$\mathfrak{b}^n_{\tau_j}(x) = \nabla B^n_{\tau_j}(x), \quad x \in \mathbb{R}^3. \tag{6.14}$$

From this we get the *lowpass filtered magnetic dipole induction* (see (5.20))

$$\mathfrak{b}^n_{\tau_j}(x) = \int_{\mathcal{B}} \mathfrak{g}^n_{\tau_j}(x,y)\ m(y)\ dV(y), \tag{6.15}$$

and the *bandpass filtered magnetic dipole induction*

$$(W\mathfrak{b})^n_{\tau_j} = \mathfrak{b}^n_{\tau_{j+1}} - \mathfrak{b}^n_{\tau_j}, \tag{6.16}$$

i. e.,

$$(W\mathfrak{b})^n_{\tau_j}(x) \;=\; \int_{\mathcal{B}} \underbrace{(\mathfrak{g}^n_{\tau_{j+1}}(x,y) - \mathfrak{g}^n_{\tau_j}(x,y))}_{=(W\mathfrak{g})^n_{\tau_j}(x,y)}\ m(y)\ dV(y). \tag{6.17}$$

For nonnegative integers j with $J_0 \leq j \leq J-1$ this allows a decorrelation of $\mathfrak{b}^n_{\tau_J}$ via the identity (see (5.22))

$$\mathfrak{b}^n_{\tau_J}(x) = \mathfrak{b}^n_{\tau_{J_0}}(x) + \sum_{j=J_0}^{J-1} (W\mathfrak{b})^n_{\tau_j}(x). \tag{6.18}$$

Therefore, for suitably sufficiently large J,

$$\mathfrak{b}(x) \simeq \mathfrak{b}^n_{\tau_J}(x) = \int_{\mathcal{B}} \mathfrak{g}^n_{\tau_J}(x,y)\ m(y)\,dV(y), \quad x \in \mathbb{R}^3, \tag{6.19}$$

holds true. Similarly, in conjunction with a suitable cubature formula for $\overline{\mathcal{B}}$, we obtain

$$\mathfrak{b}(x) \simeq \mathfrak{b}^n_{\tau_J}(x) \simeq \sum_{i=1}^{N_J} \mathfrak{g}^n_{\tau_J}(x, y_i^{N_J})\ \underbrace{w_i^{N_J}\ m(y_i^{N_J})}_{=a_i^{N_J}}, \quad x \in \mathbb{R}^3, \tag{6.20}$$

and

$$(Wb)^n_{\tau_J}(x) \simeq \sum_{i=1}^{N_J} (W\mathfrak{g})^n_{\tau_J}(x, y_i^{N_J}) \underbrace{w_i^{N_J}\ m(y_i^{N_J})}_{=a_i^{N_J}}, \quad x \in \mathbb{R}^3, \tag{6.21}$$

where $w_i^{N_J} \in \mathbb{R}$ and $y_i^{N_J} \in \mathbb{R}^3$, $i = 1, \ldots, N_J$, are the known weights and nodes, respectively. Thus, if a finite set of vector values of the field b at certain points is known, the approximate formula (6.20) yields a linear system in the unknown coefficients $a_i^{N_J}$, $i = 1, \ldots, N_J$, and consequently $m(y_i^{N_J})$, $i = 1, \ldots, N_J$.

Scalar-based variant. The field b is convolved with the scalar mollifiers $H^n_{\tau_j}(\cdot,\cdot)$, leading to

$$b^n_{\tau_j}(x) = \int_{\mathcal{B}} H^n_{\tau_j}(x,y)\ \underbrace{b(y)}_{=\nabla B(y)}\ dV(y) \tag{6.22}$$

and consequently to mollifiers

$$b^n_{\tau_j}(x) = \nabla_x \int_{\mathcal{B}} \int_{\mathcal{B}} H^n_{\tau_j}(x,y) \left(\nabla_y \frac{1}{4\pi} \frac{1}{|z-y|} \right) dV(y) \cdot m(z)\ dV(z). \tag{6.23}$$

It should be noted that the integrals

$$\iota^n_{\tau_j}(x,z) \;=\; \int_{\mathcal{B}} H^n_{\tau_j}(x,y) \underbrace{\left(\nabla_y \frac{1}{4\pi} \frac{1}{|z-y|} \right)}_{=\gamma(x,y)} dV(y) \tag{6.24}$$

for $x, z \in \mathcal{B}$, γ given as in (5.16), can be computed a priori and stored elsewhere for future use.

From (6.23) we obtain the *bandpass filtered dipole induction*

$$(WB)^n_{\tau_j}(x) = \int_{\mathcal{B}} \underbrace{(H^n_{\tau_{j+1}}(x,y) - H^n_{\tau_j}(x,y))}_{=(WH)^n_{\tau_j}(x,y)}\ b(y)\ dV(y), \tag{6.25}$$

such that for $J_0 \leq j \leq J-1$

$$b^n_{\tau_J}(x) = b^n_{\tau_{J_0}}(x) + \sum_{j=J_0}^{J-1} (WB)^n_{\tau_j}(x) \tag{6.26}$$

follows. Application of a suitable cubature formula yields

$$b^n_{\tau_J}(x) \simeq \nabla_x \sum_{i=1}^{N_J} \underbrace{\int_{\mathcal{B}} H^n_{\tau_J}(x,y) \left(\nabla_y \frac{1}{4\pi} \frac{1}{|y - y_i^{N_J}|} \right) dV(y)}_{= \iota^n_{\tau_J}(x, y_i^{N_J})} \cdot \underbrace{w_i^{N_J} m(y_i^{N_J})}_{= a_i^{N_J}}. \tag{6.27}$$

Thus, if a finite set of vector values of b is known, the cubature formula (6.27) yields a linear system in the vectorial unknowns $a_i^{N_J}$, $i = 1, \ldots, N_J$, and consequently the vector values $m(y_i^{N_J})$.

For $j = J_0, \ldots, J-1$, the constituent parts of a multiscale decorrelation are given by

$$b^n_{\tau_j}(x) = \nabla_x \sum_{i=1}^{N_J} \int_{\mathcal{B}} H^n_{\tau_j}(x,y) \left(\nabla_y \frac{1}{4\pi} \frac{1}{|y - y_i^{N_J}|} \right) dV(y) \cdot \underbrace{w_i^{N_J} m(y_i^{N_J})}_{= a_i^{N_J}} \tag{6.28}$$

and

$$(Wb)^n_{\tau_j}(x) = \nabla_x \sum_{i=1}^{N_J} \int_{\mathcal{B}} (WH)^n_{\tau_j}(x,y) \left(\nabla_y \frac{1}{4\pi} \frac{1}{|y - y_i^{N_J}|} \right) dV(y) \cdot \underbrace{w_i^{N_J} m(y_i^{N_J})}_{= a_i^{N_J}}.$$

Mollifier divergence of magnetization. We recall (cf. [54]) that for $x \in \mathcal{B}$

$$\begin{aligned} b(x) = \nabla B(x) &= \nabla_x \underbrace{\int_{\mathcal{B}} \nabla_y G(\Delta; x, y) \cdot m(y) \, dV(y)}_{= B(x)} \\ &= \nabla_x \underbrace{\left(-\nabla_x \cdot \frac{1}{4\pi} \int_{\mathcal{B}} m(y) \frac{1}{|x - y|} \, dV(y) \right)}_{= B(x)} \end{aligned} \tag{6.29}$$

holds, so that from Poisson's differential equation, e. g., under the assumption that m is Hölder-continuous, we immediately obtain

$$-\Delta B(x) = -\nabla \cdot \nabla B(x) = -\nabla \cdot b(x) = -\mu_0 (\nabla \cdot m(x)) \tag{6.30}$$

and

$$-\Delta b(x) = -\nabla(\nabla \cdot b(x)) = -\nabla(\nabla \cdot m(x)). \tag{6.31}$$

Thus, simultaneously with b, we can derive information about the macroscopic magnetic behavior of material inside $\mathcal{B}$ via the discussion of the divergence $\nabla \cdot m$ of the magnetization m. The equation

$$\nabla \cdot b = \nabla \cdot m \tag{6.32}$$

should be kept in mind, since its validity allows us to expect that the band-limited mollification of $\nabla \cdot b$ will also provide valuable geologic information for the test area under consideration:

$$\begin{aligned} b(x) = \nabla B(x) &= \nabla_x \underbrace{\int_{\mathcal{B}} \nabla_y G(x,y) \cdot m(y)\, dV(y)}_{=B(x)} \\ &\simeq b^n_{\tau_J}(x) = \nabla B^n_{\tau_J}(x) \\ &= \nabla_x \underbrace{\left(-\nabla_x \cdot \int_{\mathcal{B}} m(y)\; G_{\tau_J}(x,y)\, dV(y) \right)}_{=B^n_{\tau_J}(x)} . \end{aligned} \tag{6.33}$$

For $j \in \mathbb{N}_0$, the lowpass filtered mollifier version of the divergence of the magnetization can be introduced by

$$M^n_{\tau_j} = -\Delta B^n_{\tau_j}. \tag{6.34}$$

In the usual way, we can define the bandpass filtered mollifier version $(WM)^n_{\tau_j}$ of the divergence M of the magnetization by

$$\begin{aligned} M^n_{\tau_j}(x) = -\Delta B^n_{\tau_j}(x) &= -\nabla \cdot b^n_{\tau_j}(x) \\ &= -\nabla_x \cdot \underbrace{\int_{\mathcal{B}} m(y)\; H^n_{\tau_j}(x,y)\, dV(y)}_{=m^n_{\tau_j}(x)} \end{aligned} \tag{6.35}$$

and

$$\begin{aligned} (WM)^n_{\tau_j}(x) = -\Delta (WB)^n_{\tau_j}(x) &= -\nabla \cdot (Wb)^n_{\tau_j}(x) \\ &= -\nabla_x \cdot \underbrace{\int_{\mathcal{B}} m(y)\; (WH)^n_{\tau_j}(x,y)\, dV(y)}_{=(Wm)^n_{\tau_j}(x)} . \end{aligned} \tag{6.36}$$

Thus, the relations

$$\begin{aligned} \nabla \cdot b^n_{\tau_j}(x) &= \nabla_x \cdot \int_{\mathcal{B}} m(y)\; H^n_{\tau_j}(x,y)\, dV(y) \\ &= \nabla \cdot\; m^n_{\tau_j} \end{aligned} \tag{6.37}$$

and

$$\nabla \cdot (Wb)^n_{\tau_j}(x) = \nabla_x \cdot \int_{\mathcal{B}} m(y)\ (WH)^n_{\tau_j}(x,y)\ dV(y) \tag{6.38}$$
$$= \nabla \cdot\ (Wm)^n_{\tau_j},$$

hold true, where $(Wm)^n_{\tau_j}$ is given as usual by

$$(Wm)^n_{\tau_j} = m^n_{\tau_{j+1}} - m_{\tau^n_j}. \tag{6.39}$$

For suitable sufficiently large J we now obtain the telescope sum

$$M^n_{\tau_J}(x) = M^n_{\tau_{J_0}}(x) + \sum_{j=J_0}^{J-1} (WM)^n_{\tau_j}(x), \tag{6.40}$$

so that we get the following approximate formula

$$\nabla \cdot m \simeq \nabla \cdot m^n_{\tau_J} \tag{6.41}$$
$$= \nabla \cdot m^n_{\tau_{J_0}} + \sum_{j=J_0}^{J-1} \left(\nabla \cdot (Wm)^n_{\tau_j} \right).$$

Based on the measurement of the vector magnetic field b, the inversion is done by a suitable cubature formula analogous to the gravimetric case, i. e., Gaussian-averaged Euler-Poisson summation formulas (cf. [58] and for the numerical implementation [28]). The data-related difficulty with the linear system of equations to be solved in this process is that it is usually massively underdetermined. For this reason, preinformation about the magnetization is essential. This can be added to the original linear equation system in the form of additional equations.

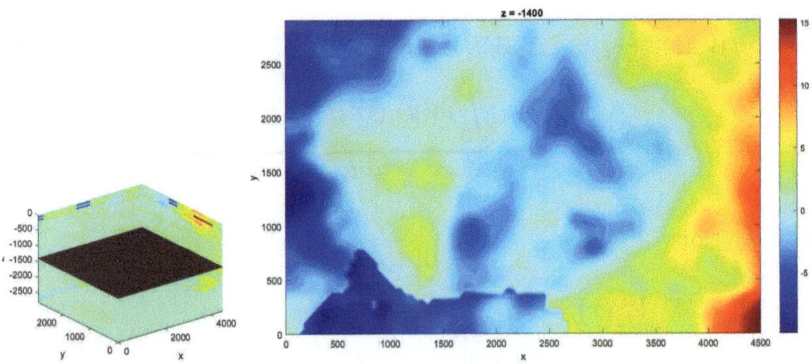

Fig. 6.3 Intersection of the divergence M of the magnetization m.

For numerical realizations of the developed decorrelation methods using the example of magnetometry, we interpret an available density data set of the Traunreut area in Figure 6.1 as divergence M of the magnetization for test purposes (cf. Figure 6.3). This approach provides a test dataset of the Traunreut area varying in all spatial directions and establishes a concrete link to geologic exploration. The test variable magnetization m (cf. Figure 6.4) for Traunreut is then arrived at by, for example, discretizing the divergence operator in the equation by a finite difference scheme.

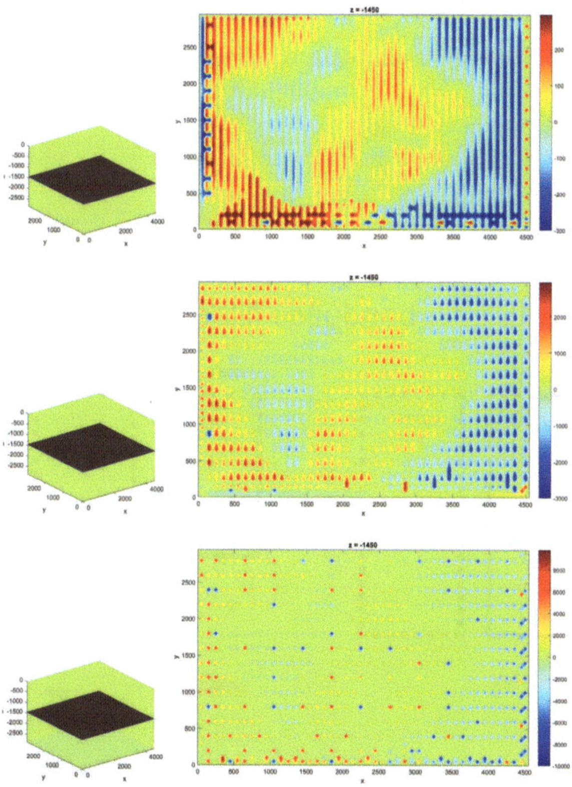

Fig. 6.4 Intersection of magnetization m in its three vector components (x_1-component top, x_2-component middle, x_3-component bottom).

The illustrations below show a graphical illustration of the developed multiscale method.

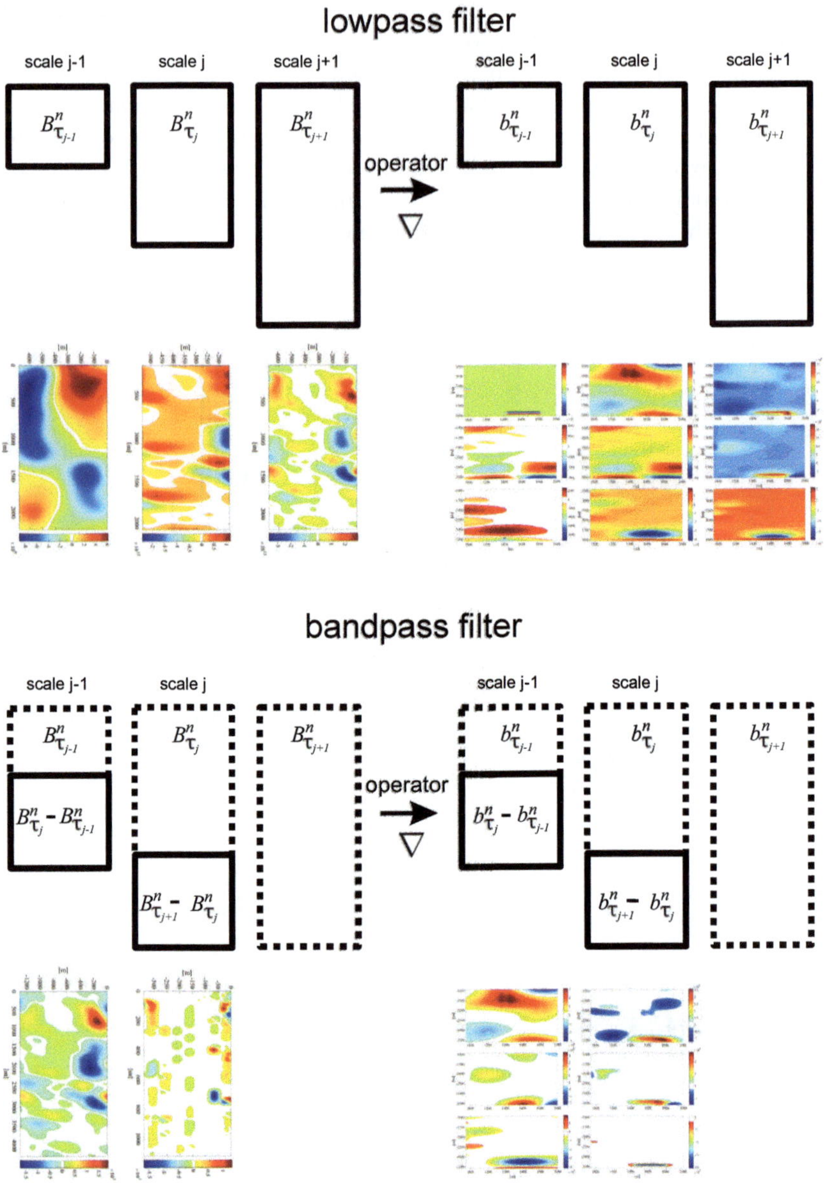

Fig. 6.5 Schematic correlation of decorrelations of dipole potential and dipole induction using the example of Traunreut, Bavarian Molasse Basin.

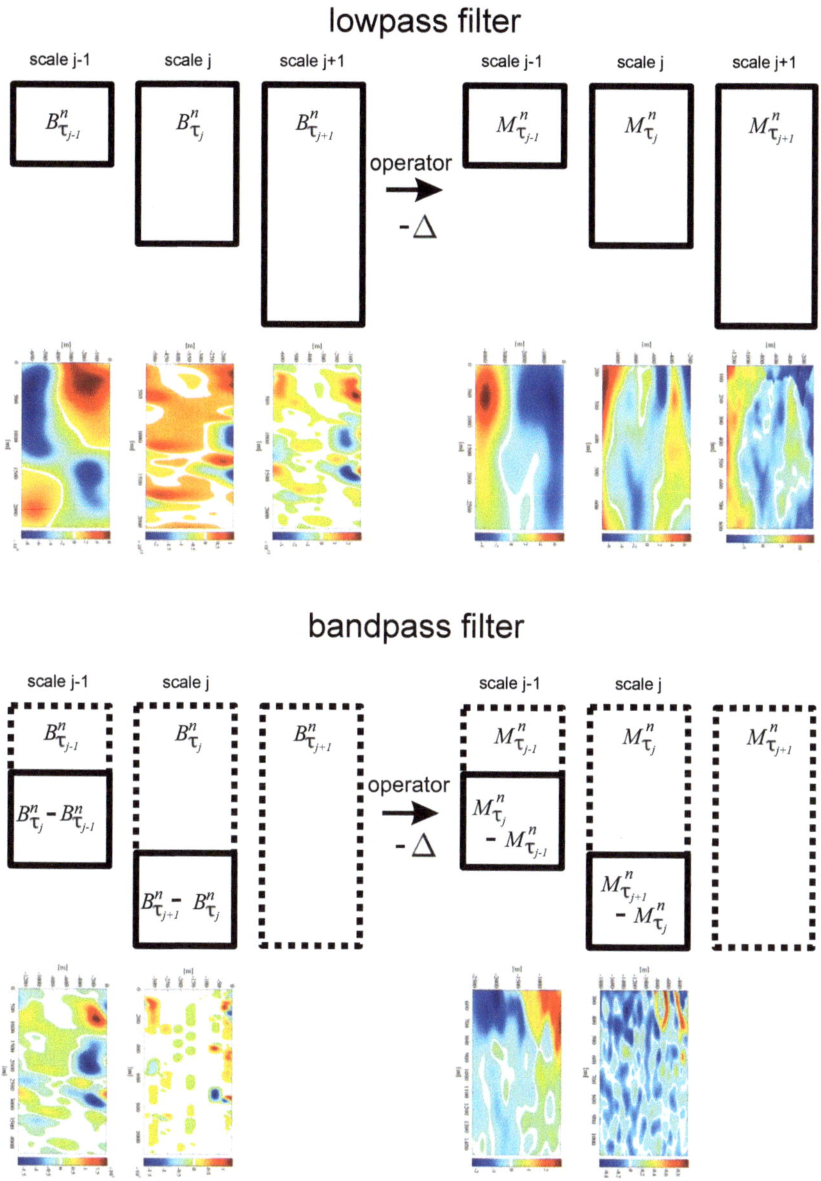

Fig. 6.6 Schematic correlation of decorrelations of dipole potential and divergence of magnetization using the example of Traunreut, Bavarian Molasse Basin.

Figure 6.5 shows the relationship between the decomposition of $B^n_{\tau_J}$ and $b^n_{\tau_J}$. It turns out that the decomposition of the dipole potential $B^n_{\tau_J}$ provides little useful information about the geology because of its smoothness (Figure 6.5, top left). However, if we proceed to multiscale decomposition in the form of bandpass-filtered data to scale j, $B^n_{\tau_j} - B^n_{\tau_{j-1}}$, structural information, as found in the magnetization m, becomes visible (Figure 6.5, bottom left). By applying the gradient, we obtain a decomposition of $b^n_{\tau_J}$ that is still smooth and does not give any information about the geology (Figure 6.5, top right). However, if we take the difference $b^n_{\tau_j} - b^n_{\tau_{j-1}}$ (Figure 6.5, bottom right), we obtain vectorial geologic information comparable in the sum of its vector components to the negative magnetic divergence $-\nabla \cdot m^n_{\tau_{j-1}}$ (Figure 6.6, top right). In Figure 6.6 we compare the decomposition of $B^n_{\tau_J}$ with the decomposition of the magnetic divergence $M = -\nabla \cdot m$. The left side corresponds to Figure 6.5. However, instead of the gradient, we apply the negative Laplace operator to obtain the decomposition of the magnetic divergence on the right-hand side. Moreover, from the difference $-\nabla \cdot m^n_{\tau_j} + \nabla \cdot m^n_{\tau_{j-1}}$ the local transition zones of $-\nabla \cdot m$ become available.

All in all, we have a powerful tool at hand that provides geologic information about the magnetization of a study area.

6.4 Gravimetric spline wavelet mollifier methodology

Another method of reservoir modeling is made possible by the use of mollifier repro kernel spline wavelets. The method has the advantage that a uniform distribution of the data points does not have to be given necessarily, but arbitrarily distributed data point systems can be computed directly. The evaluation can be carried out without a "calculation from top to bottom" for the entire subsurface, whereby there are restrictions with regard to the depth because of the support width of the mollifer, i. e., the narrower the filter, i. e. the finer the detailed structures, the less deep the calculation can be carried out. Thus, this method also offers the possibility to model and represent the interfaces and layer boundaries in the subsurface, locally for each subarea given the appropriate available data material. A completion of the data material by best approximation is also possible.

The basic idea of the method of repro splines shall be explained for the sake of simplicity by the example of gravimetry. It is based on the insight that the function space Y of all Newton volume potentials

$$Y = \{P \, : \, P = \int_{\mathcal{B}} \underbrace{\frac{1}{4\pi} \frac{1}{|\cdot - y|}}_{=G(\cdot, y)} \rho(y) \, dV(y), \; \rho \in L^2(\mathcal{B})\} \tag{6.42}$$

for square-integrable densities ρ in $\mathcal{B}$ is an infinite-dimensional Hilbert space with repro spline structure

$$< K_Y(x,\cdot),P >_Y = P(x), \quad x \in \mathbb{R}^3, \; P \in Y, \tag{6.43}$$

with the reproducing kernel

$$K_Y(x,y) = \int_{\mathcal{B}} \underbrace{\frac{1}{4\pi}\frac{1}{|x-z|}}_{=G(x,z)} \underbrace{\frac{1}{4\pi}\frac{1}{|y-z|}}_{=G(y,z)} \, dV(z), \quad x,y \in \mathbb{R}^3. \tag{6.44}$$

The repro structure implies that a convolution of the potential P with the kernel $K_Y(\cdot,\cdot)$ (in the Y-topology) leads again to P. Methodically, one saves filtering by integration, but at the price of the transition from the $L^2(\mathcal{B})$ topology to the Y topology. Introducing splines of the form (cf. [51])

$$S(x) = \sum_{i=1}^{N} a_i^N \; K_Y(x,x_i), \quad x,x_i \in \mathbb{R}^3, \tag{6.45}$$

which can be formed using the reproducing kernel (6.44), then leads to a solution of a system of equations whose coefficient matrix has a highly advantageous structure as the Gram matrix

$$K_Y(x,x_i) = < K_Y(x,\cdot), K_Y(\cdot,x_i) >_Y. \tag{6.46}$$

The kernel function $K_Y(x,y)$ generated by applying the Newton potential operator to the Newton kernel allows to perform spline interpolation due to the repro kernel property: Starting from N arbitrarily distributed measurement points x_i in the measurement domain with associated potential data $V(x_i)$, we search for an approximate solution (spline interpolant)

$$S_N^V \approx V \tag{6.47}$$

of the potential V. The spline solution thus satisfies the interpolation property

$$S_N^V(x_i) = \sum_{k=1}^{N} a_k^N \; K_Y(x_k,x_i) = V(x_i), \quad i = 1,\dots,N. \tag{6.48}$$

An advantage of the spline method using repro kernels is the possibility to replace the Hilbert space Y by a mollifier variant $Y^n_{\tau_j}$:

$$Y^n_{\tau_j} = \{P \,:\, P = \int_{\mathcal{B}} G^n_{\tau_j}(x,y) \; F(y) \; dV(y), \;\; F \in L^2(\mathcal{B})\}. \tag{6.49}$$

All the above statements remain valid if instead of $G(\cdot,\cdot)$ we write the mollifier counterpart $G^n_{\tau_j}(\cdot,\cdot)$. In this way, depending on the scale j, structures in the dataset can be made visible and band structures can be prepared out. Even more, noisy datasets can be treated with appropriate smoothing methods or combined interpolation/smoothing methods (cf. [54, 80]).

The mollifier spline method can alternatively be implemented in the following way:

(G1) Starting point is the mollifier-Newton integral

$$V_{\tau_j}^n(x) = \int_{\mathcal{B}} G_{\tau_j}^n(x,z)\ \rho(z)\ dV(z), \quad x \in \mathbb{R}^3, \tag{6.50}$$

for an $L^2(\mathcal{B})$-integrable mass density distribution ρ.

(G2) The mass density distribution is assumed to be a linear combination

$$\rho(x) = \sum_{i=1}^{N} a_i^N\ G_{\tau_j}^n(x,x_i) \tag{6.51}$$

at the data points $x_i \in \mathbb{R}^3$, $i = 1,...,N$ (so that it is non-harmonic by its nature).

(G3) Substituting (6.51) into (6.50) then yields

$$V_{\tau_j}^n(x) = \sum_{i=1}^{N} a_i^N \int_{\mathcal{B}} G_{\tau_j}^n(x,z)\ G_{\tau_j^n}(z,x_i)\ dV(z). \tag{6.52}$$

(G4) The requirement $V_{\tau_j}^n(x_k) = V(x_k)$, $k = 1,...,N$, makes the connection to the $Y_{\tau_j}^n$-spline interpolation via the linear equation system

$$V(x_k) = \sum_{i=1}^{N} a_i^N\ K_{Y_{\tau_j}^n}(x_k,x_i) \tag{6.53}$$

with

$$K_{Y_{\tau_j}^n}(x,y) = \int_{\mathcal{B}} G_{\tau_j}^n(x,z)\ G_{\tau_j}^n(y,z)\ dV(z), \quad x,y \in \mathbb{R}^3. \tag{6.54}$$

The method can also be applied to functionals of the gravitational field, such as radial derivatives (also of higher order), inclined-axial derivatives or negative Laplace derivatives, in all points $x_i \in \mathbb{R}$. In particular, terrestrial gravimetric data and even gravimetric borehole data can find input as derivative functionals. This formally leads to a combined interpolation/smoothing mollifier approximation. Furthermore, in the mollifier context, realistic information about potential and density can be treated in combination by means of an interpolation/smoothing spline mechanism (more details can be found in [54]).

The following Figures now show results of the procedure for the test dataset Traunreut in the Bavarian Molasse basin (see also Section 6). For this dataset, the density in the subsurface is known as a model. Synthetic (ideal) measurement data were calculated in 570 uniformly distributed measurement points (see Figure 6.7).

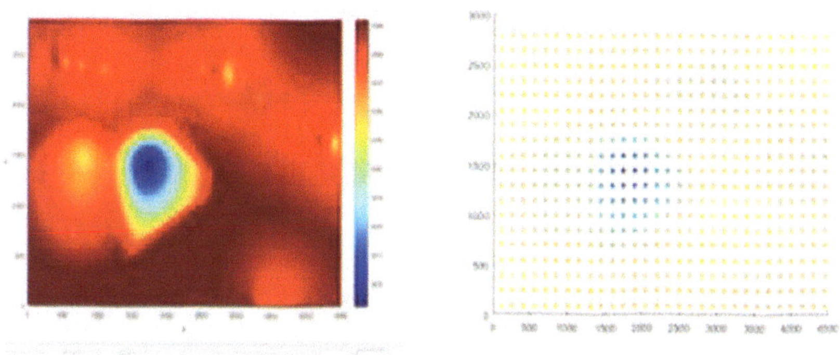

Fig. 6.7 Density data from Traunreut, Bavarian Molasse Basin, at the surface (left) and gravimetric measurements calculated from the entire density dataset at 570 measurement points at the surface (right).

In Figure 6.8, on the left, one can see the lowpass filtering of the density calculated from gravimetric potential data for different filter widths, on the right the corresponding bandpass filtering is shown. The evaluations are performed at the surface. As the scale increases, i. e., the filter width decreases, finer and finer details become visible. Due to the selected data density of 150 m, finer details are not detectable.

The method of decorrelation using reproducing kernel spline mollifier wavelets allows evaluation for arbitrary areas of the subsurface, where the filter width imposes a constraint on depth, i. e., the narrower the filter width, the shallower the maximum depth that can be represented.

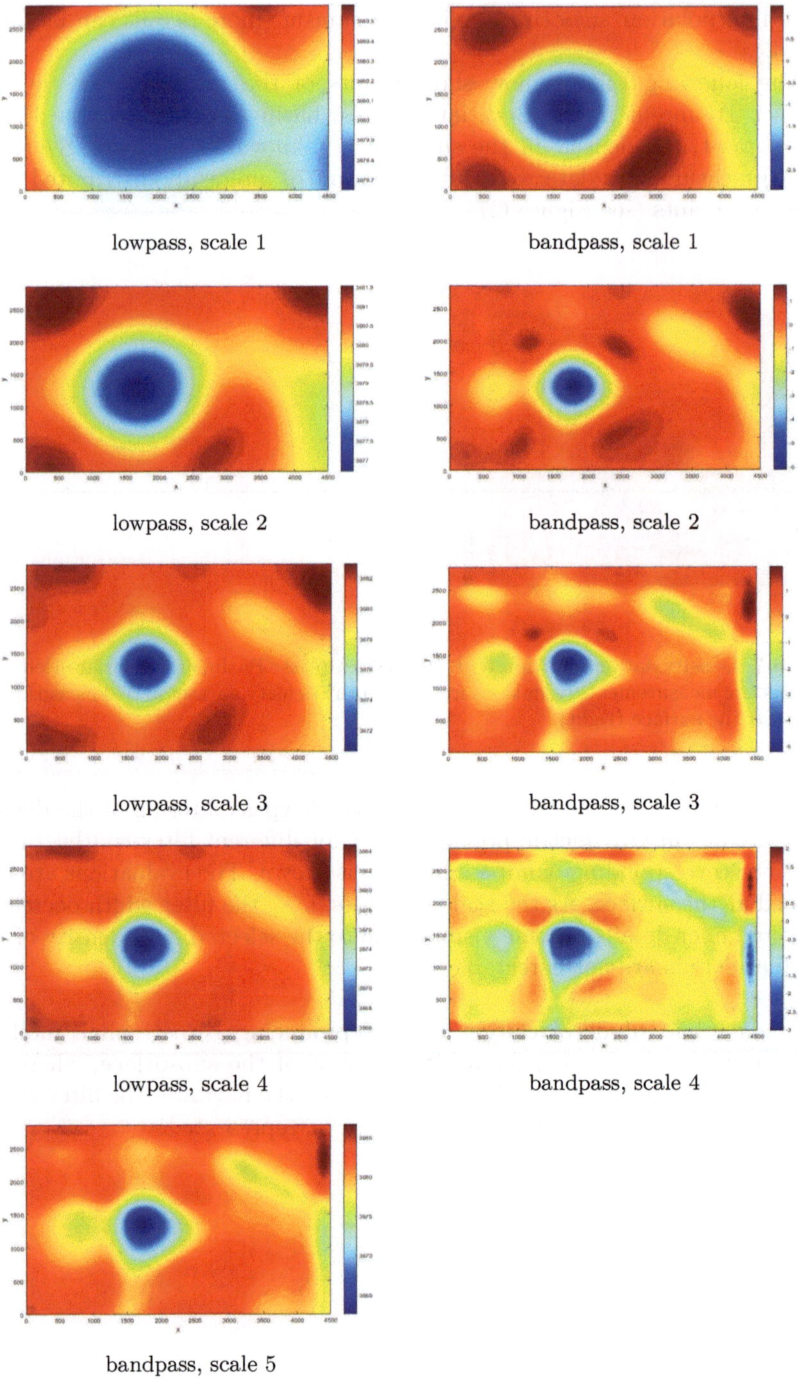

Fig. 6.8 Decorrelation using the example of Traunreut, Bavarian Molasse Basin, evaluation at the surface based on gravimetric data at the surface (see Figure 6.7, right).

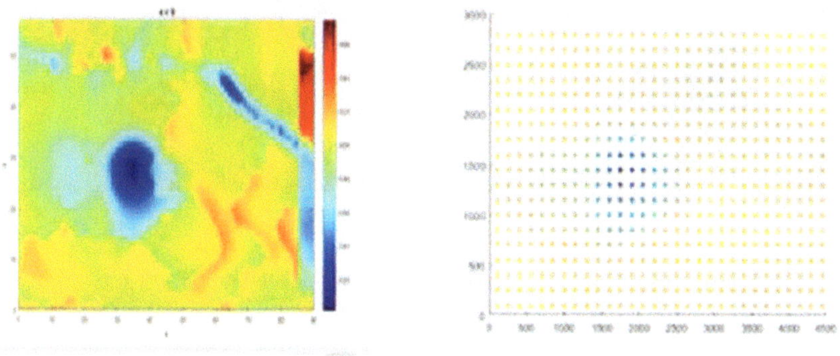

Fig. 6.9 Density data from Traunreut, Bavarian Molasse Basin, at a depth of about 400 m (left) and gravimetric measurements calculated from the entire density dataset at 570 measurement points at the surface (right).

In Figure 6.9 the density using the example of Traunreut, Bavarian Molasse Basin, in the subsurface at a depth of about 400 m is shown on the left, on the right the already known synthetic measurement data in 570 evenly distributed measurement points (cf. also Figure 6.7).

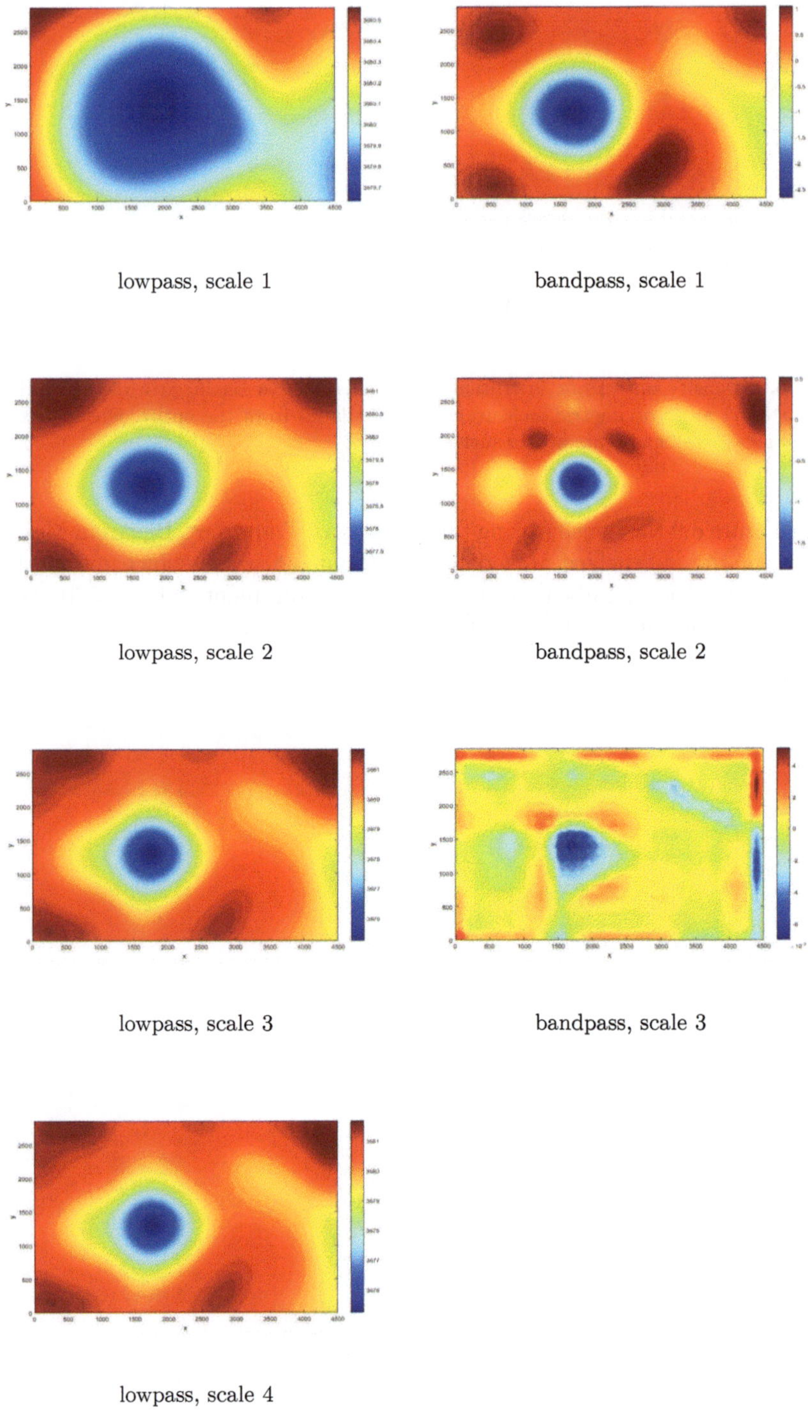

Fig. 6.10 Decorrelation using the example of Traunreut, Bavarian Molasse Basin, evaluation at a depth of about 400 m based on gravimetric data at the surface (see Figure 6.7, right).

Figure 6.10 shows the evaluation of decorrelations with reproducing kernel spline wavelets at the depth of about 400 m. For better interpretability of the results, one needs information from the interior, such as borehole data. The method allows the direct integration of this information into the calculations.

6.5 Magnetometric spline wavelet mollifier methodology

The points (G1)–(G4) from Section 6.4 now also allow the transition to magnetometry in the following way:

(M1) The starting point is the mollifier dipole integral

$$B^n_{\tau_j} = \int_{\mathcal{B}} \gamma^n_{\tau_j}(x,y) \cdot m(y)\ dV(y), \quad x \in \mathbb{R}^3, \tag{6.55}$$

for an $l^2(\mathcal{B})$-integrable magnetization distribution m.

(M2) The magnetization distribution is taken to the data points $x_i \in \mathbb{R}^3$, $i = 1, ..., N$, understood as a spline linear combination

$$m(x) = \sum_{i=1}^{N} a_i^N\ \gamma^n_{\tau_j}(x, x_i). \tag{6.56}$$

(M3) Substituting (6.56) into (6.55) we obtain

$$B^n_{\tau_j}(x) = \sum_{i=1}^{N} a_i^N \int_{\mathcal{B}} \gamma^n_{\tau_j}(x,z) \cdot \gamma^n_{\tau_j}(z, x_i)\ dV(z). \tag{6.57}$$

(M4) The requirement $B^n_{\tau_j}(x_k) = B(x_k)$, $k = 1, ..., N$, then makes the connection to the $Y^n_{\tau_j}$-magnetometry spline interpolation via the linear system of equations

$$B(x_k) = \sum_{i=1}^{N} a_i^N\ k_{Y^n_{\tau_j}}(x_k, x_i) \tag{6.58}$$

with the repro kernel

$$k_{Y^n_{\tau_j}}(x,y) = \int_{\mathcal{B}} \gamma^n_{\tau_j}(x,z) \cdot \gamma^n_{\tau_j}(y,z)\ dV(z), \quad x, y \in \mathbb{R}^3. \tag{6.59}$$

For a detailed specification of the $Y^{n}_{\tau_j}$-magnetometry repro structure, we refer to the monograph [54].

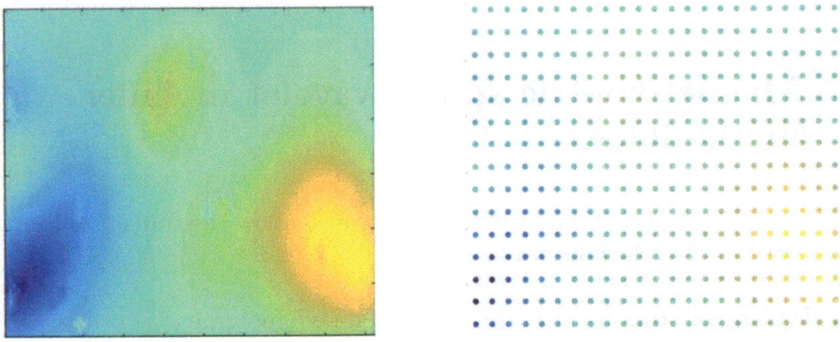

Fig. 6.11 On the left is the plot of magnetic potential data calculated from density data from Traunreut, Bavarian Molasse Basin, and on the right is the measurement data extracted from it at 345 grid-like distributed measurement points.

We made a test of this decorrelation method on the test dataset Traunreut, Bavarian Molasse basin (see also Section 6). Figure 6.11 shows the input data at 345 uniformly distributed measurement points. The mollifier method of vector repro kernel spline wavelets is applied to these data (see Figures 6.12 to 6.15).

Figures 6.12 and 6.13 show a decorrelation with repro kernel spline wavelets computed at the surface.

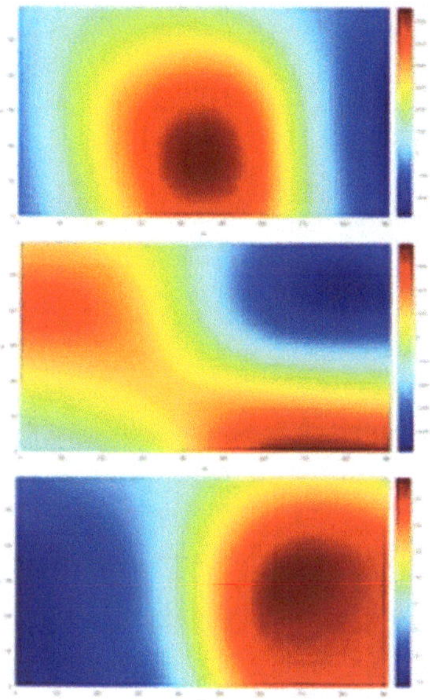

lowpass, scale 1

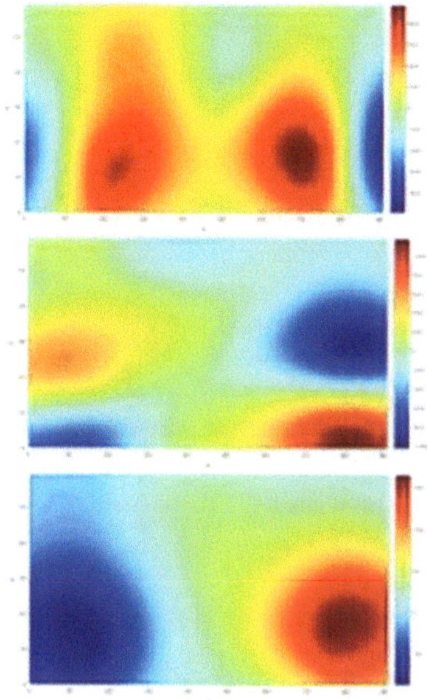

bandpass, scale 1

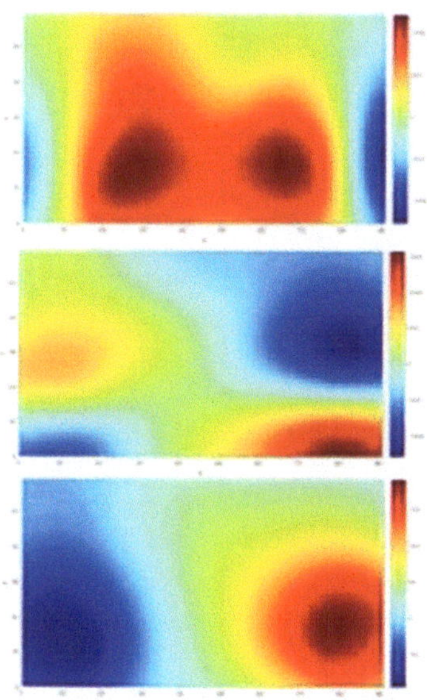

lowpass, scale 2

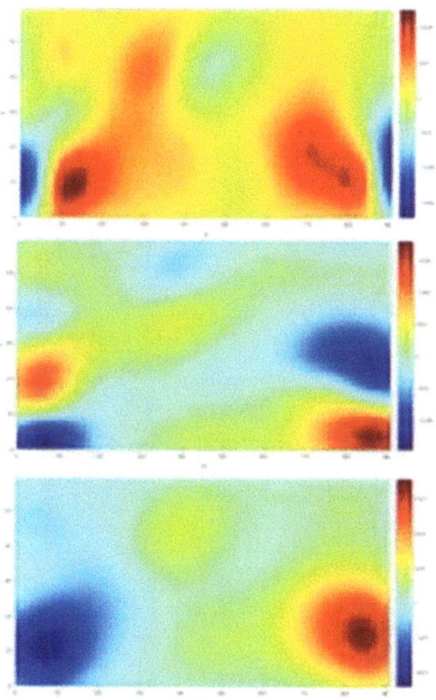

bandpass, scale 2

Fig. 6.12 Plot of the lowpass and bandpass components (from top to bottom, the x_1, x_2, and x_3 components, respectively). Evaluation at the surface by decorrelation with vector repro kernel spline wavelets based on data of the magnetic potential of Traunreut, Bavarian Molasse Basin, at the surface (see Figure 6.11).

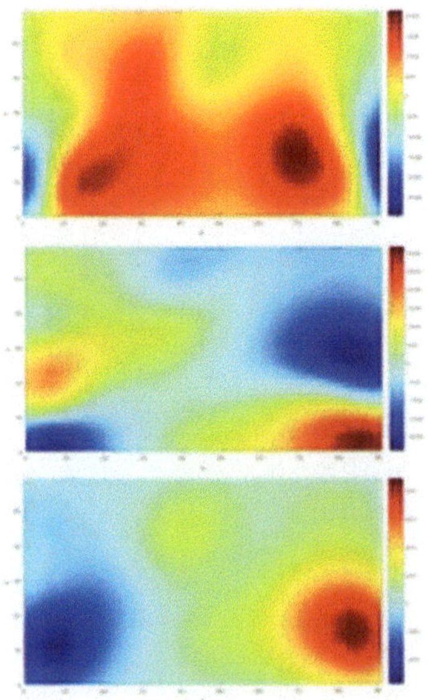

lowpass, scale 3

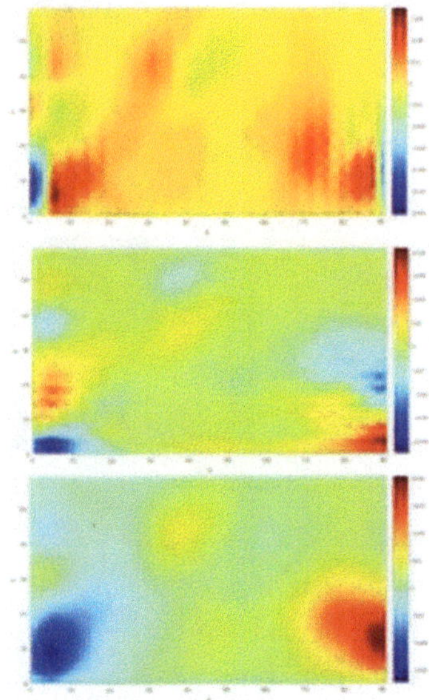

bandpass, scale 3

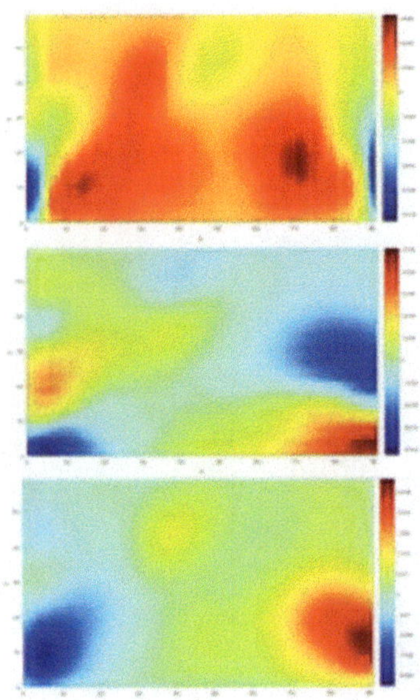

lowpass, scale 4

Fig. 6.13 Plot of the lowpass and bandpass components (from top to bottom, the x_1, x_2, and x_3 components, respectively). Evaluation at the surface by decorrelation with vector repro kernel spline wavelets based on data of the magnetic potential of Traunreut, Bavarian Molasse Basin, at the surface (see Figure 6.11), continued.

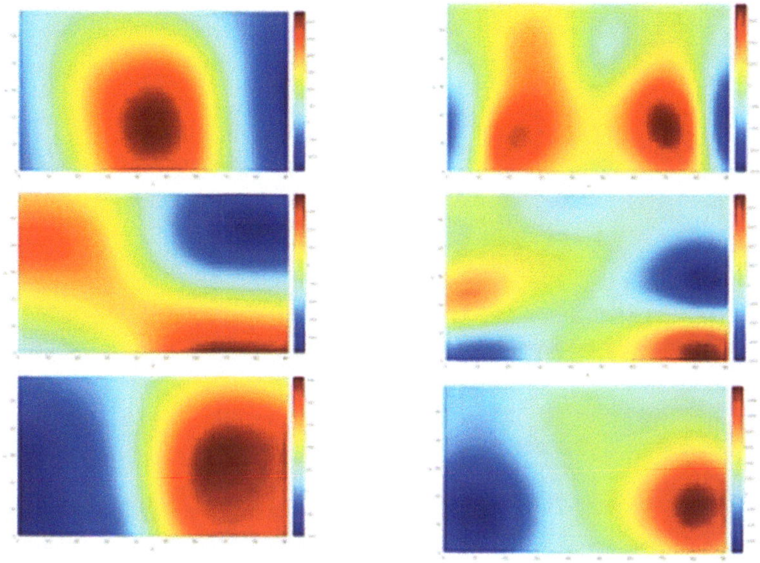

lowpass, scale 1 bandpass, scale 1

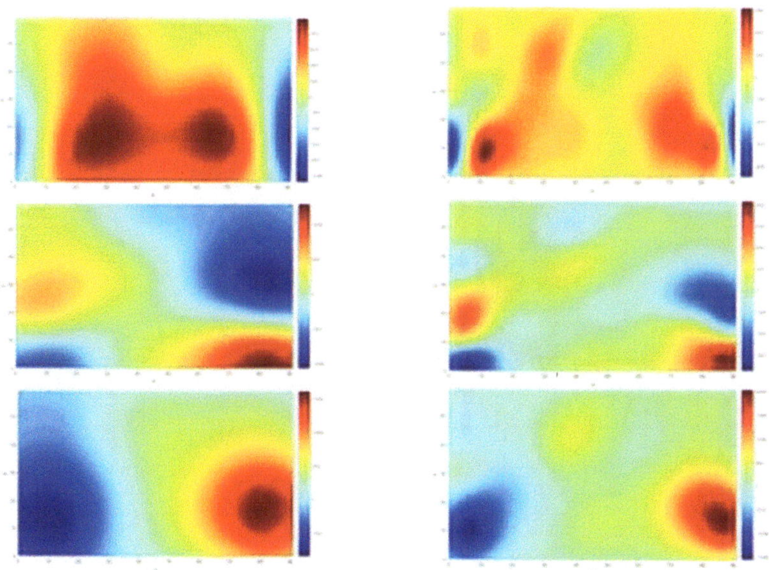

lowpass, scale 2 bandpass, scale 2

Fig. 6.14 Plot of the lowpass and bandpass components (from top to bottom, the x_1, x_2, and x_3 components, respectively). Evaluation at a depth of about 400 m by decorrelation with vector repro kernel spline wavelets based on data of the magnetic potential of Traunreut, Bavarian Molasse Basin, at the surface (see Figure 6.11).

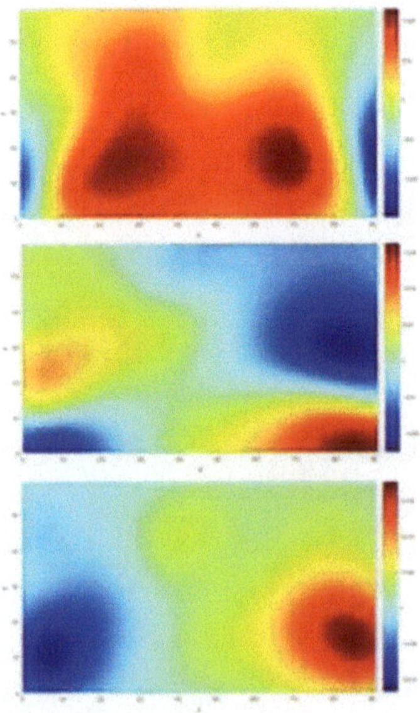

lowpass, scale 3

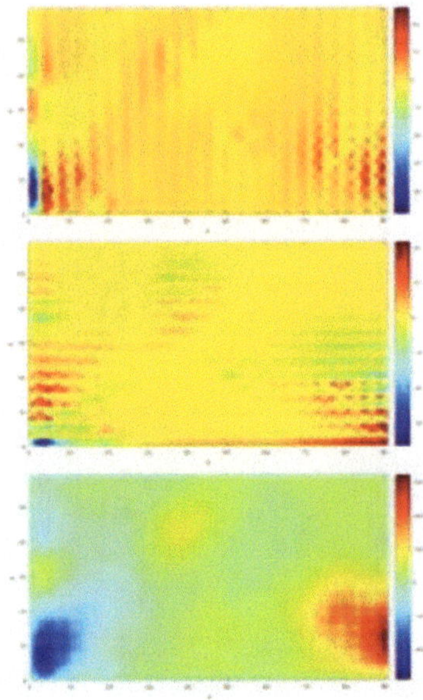

bandpass, scale 3

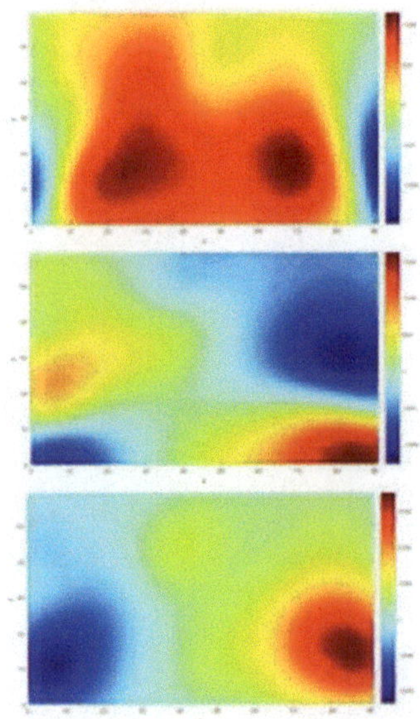

lowpass, scale 4

Fig. 6.15 Plot of the lowpass and bandpass components (from top to bottom, the x_1, x_2, and x_3 components, respectively). Evaluation at a depth of about 400 m by decorrelation with vector repro kernel spline wavelets based on data of the magnetic potential of Traunreut, Bavarian Molasse Basin, at the surface (see Figure 6.11), continued.

As already mentioned, the transition to depth is feasible with this method without further methodological effort if suitable datasets are available. However, due to the band structure of the repro kernel spline wavelets used, the maximum depth at which analysis using surface data is possible depends on the filter width selected. Figures 6.14 and 6.15 show a decorrelation calculated at a depth of about 400 m.

The dipole character of the reproducing kernel spline wavelets used is reflected in the decorrelations and must be included in a geologic interpretation of the results. In all three components of the decorrelation, detailed structures resulting from the underlying magnetic potential can be seen, whereby the third component (x_3-direction) provides the best results due to the dipole character in the selected horizontally oriented representation.

Chapter 7

Data collection in the Saarland test areas

Abstract:

This chapter deals with the collection of the data that were gravimetrically and/or magnetometrically required for the application of the multiscale mollifier potential methods in areas of the Saarland. For this purpose, first of all a listing of the already available Saarland data material was made and documented with respect to their spatial/temporal specifics in numerical implementations. The measurement instruments used for an additional data generation, their handling and the resulting measurement accuracies are recorded as well as the amount of work required to guarantee a certain data density in the test areas.

7.1 Test area Saar-Nahe basin

Brief Political Genesis of the German State Saarland. The state Saarland lies in the southwest of Germany, on the French border (cf. Figure 7.1) in the neighbourhood of the German state of Rhineland-Palatinate. Coal mining and iron industry were the most important branches of the economy. The formation of the Saarland and its changeable political history are strongly related to the geologic conditions of its underground (cf. [160]). Between 1920 and 1957 the affiliation of the Saarland changed between France and Germany. As part of the Paris Treaties, the German Chancellor K. Adenauer and the French Prime Minister P. Mendès France at that time decided on the so-called Saar Statute in October 1954. The Saar Statute provided for the Saarland to be supervised by a European Commissioner who in turn would be

W. Freeden, H. Nutz, *Exploratory Potential Methods in Geothermal Power Generation*, Lecture Notes in Geosystems Mathematics and Computing,
https://doi.org/10.1007/978-3-031-54412-5_7

responsible to the Council of Ministers of the Western European Union until a peace treaty was signed. After the Saar population had rejected the Saar Statute in a referendum in 1955, the Saarland was politically reintegrated with West Germany as a federal state in accordance with the German-French Saar Agreement in 1957. The complete reunification did not take place until July 6, 1959, when the economic annexation to Germany also took place and the D-Mark replaced the Franc(see also [191, 192, 193]).

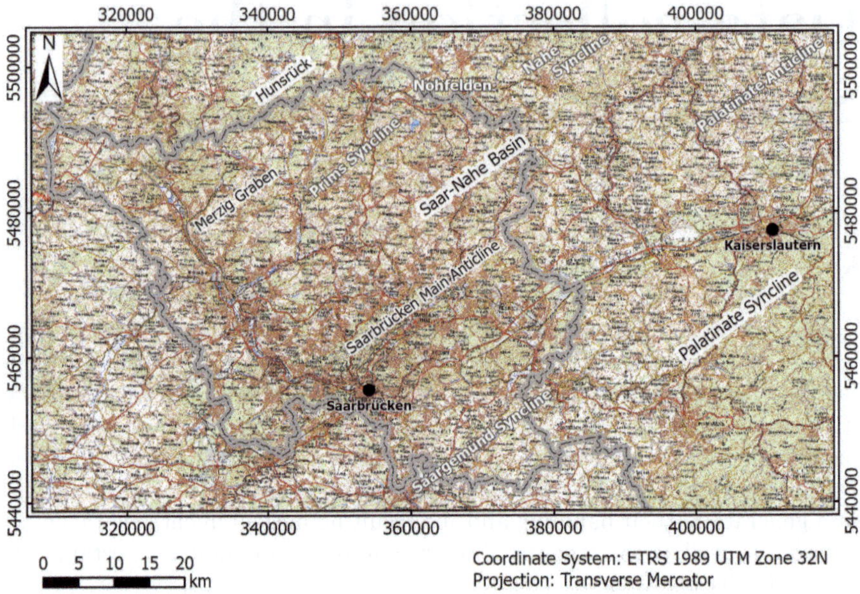

Fig. 7.1 Map of the Saarland/Palatinate reference area.

Brief Political Genesis of the German State Rhineland-Palatinate. According to the Treaty of Campo Formio 1797, the lands to the west of the Rhine were incorporated into the French territories, the individual states were dissolved, and religious holdings were secularized. In 1815 Rhineland became a part of the newly founded German Confederation, and the region was divided by the Congress of Vienna among Prussia (Preußen), Bavaria (Bayern), Hessen (Darmstadt), and Nassau. After World War II the region was again divided by order of the U.S., French, and British military governments, and the present-day German state of Rhineland-Palatinate (German: Rheinland-Pfalz) was created.

Rhineland-Palatinate (cf. Figure 7.2) is bordered by the German states of North Rhine–Westphalia to the north, Hessen to the east, Baden-Württemberg to the southeast, and Saarland to the southwest and by France, Luxembourg, and Belgium to the south and west. The state is part of the Rhine

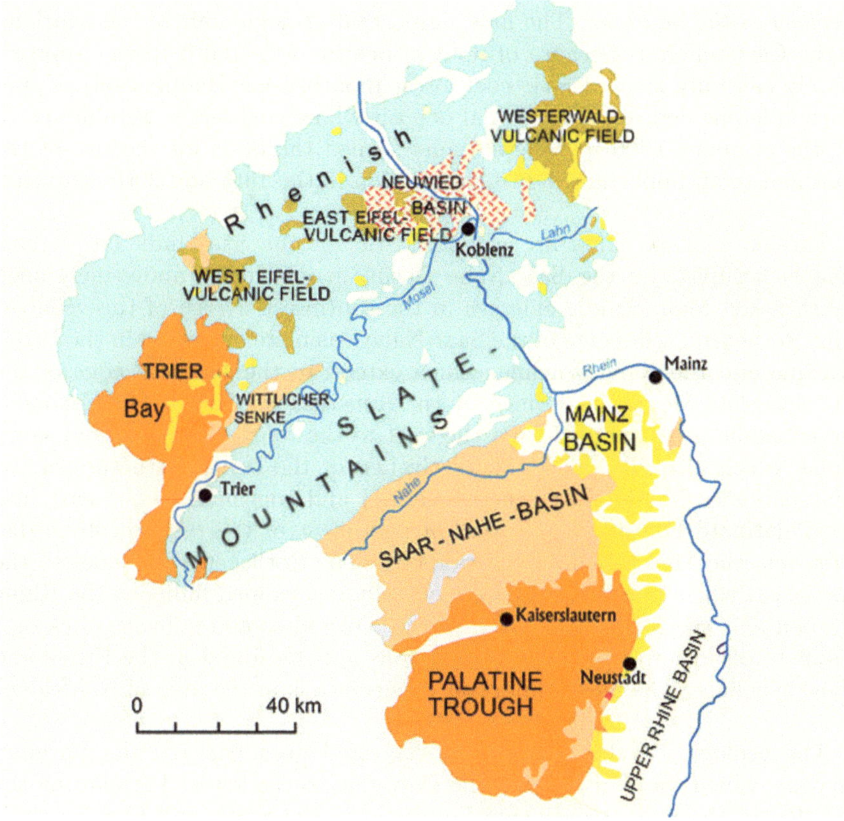

Fig. 7.2 Landscapes in Rhineland-Palatinate due to "Landeszentrale Politische Bildung", http://rlp100seiten.de/land-und-landschaften (in modified form).

Massif. The Rhine River crosses the region from the southeast to the northwest and receives the Moselle (German: Mosel) River from the southwest and the Lahn River from the northeast. The rivers effectively create four distinct regions in the northern two-thirds of the state: the Hunsrück, the Taunus, the Eifel, and the Westerwald. Rising south of the Moselle River, the Hunsrück Plateau forms the most southerly section of the uplands. The Rhine River divides the Hunsrück from the very similar Taunus Plateau to the east. North and west of the Moselle and Rhine valleys, the Neuwied Basin and Maifeld form a transition to the wooded upland of the Eifel. The Eifel landscape is dotted with volcanic craters, most of which are filled by small lakes. To the northeast is the rolling plateau of the Westerwald, which is composed mainly of slate with some higher, rounded basalt hills. The Westerwald is bounded to the north by the highly industrialized Sieg River valley (for more details about Rhineland-Palatinate (Rheinland-Pfalz) see, e. g. [187, 190].

Geology in the Saarland. The most important geologic unit of the Saarland is the Carboniferous because of its coal bearing (cf. [160]). In the approximately centrally located Saar coal forest (Saarbrücker Hauptsattel), Upper Carboniferous deposits form a coal deposit whose coal seams were mined to depths of about 1700 m. These seams formed the basis for the rise of the Saarland to an important industrial location in the 19th and 20th centuries.

Various geologic units are characteristic for the Saarland. The largest area is occupied by the Saar-Nahe Basin, in which the above-mentioned Saarbrücken Main Saddle emerges in the southwest. North of this, concordant Rotliegend sediments of the Saar-Nahe Basin are deposited in the Prims syncline and the Nahe syncline, which extend to the southern edge of the Rhenish Slate Mountains formed by the Hunsrück (cf. [160]). The Carboniferous Saddle is surrounded from the west to the southeast by the red sandstone, which also fills the Merziger Graben or the trough structure of the Saargemünder Mulde (Zweibrücker Mulde) and continues to the east into the Palatinate Forest. To the west or southwest of the red sandstone distribution, the Muschelkalk deposits follow. The Rotliegend continues to the northeast, which finally finds its tectonically determined limits at the Rhine Graben and the Mainz Basin (cf. [160]). Volcanites and subvolcanites of a distinct volcanism of Permo-Carboniferous age are found in the Prims and Nahe synclines. The most important occurrence is in the area of Nohfelden.

The geologic development of the Saarland goes back to the Variscan orogeny, which was caused from the Devonian to the lowest Permian by the collision of the major continents Laurussia in the north and Gondwana in the south with the intervening terranes and terrane complexes (Avalonia, Armorika, Norika) (cf. [121]). According to the outline of [112] and later supplemented in numerous works (see, e. g., [38, 49]), in the area of the Saarland the Variscan deformed basement is mainly determined by the Middle German Crystalline Zone in the southeast, which is seen as a tectonic structural element at the northwest margin of the Saxothuringian, and the Rhenoherzynian in the northwest. The narrow highly metamorphic zone at the southern margin of the Rhenoherzynic is defined as the Northern Phyllic Zone. This is exposed in the Saarland in the Düppenweiler floe (see [114, 182]). At the northern edge of Saarland, in the Hunsrück, the units of the Rhenoherzynic are found with low-metamorphic Devonian series of claystones, siltstones and quartzitic sandstones (see [7, 166]).

Towards the end of the Variscan orogeny (Visé/Namur), numerous intramontane Molasse basins formed mostly along the older tectonic structures. Among these, the Saar-Nahe basin is one of the largest, which developed as a half-graben along the synsedimentary active southwest-northeast striking listric and south dipping Hunsrück-Südrand fault (cf. [166]). This fault zone also separates the two major geotectonic units, the Central German Crystalline Zone and the Rhenoherzynic. Different opinions are held about the nature of this fault, which strongly influenced the evolution of the Saar-Nahe Basin (see, e. g., [1, 6, 98, 110, 111]).

During the orogeny the synsedimentary active longitudinal and transverse faults led to the formation of grabens and horsts and parallel to the longitudinal basin axis to anticlinal and synclinal structures (see [97, 166]). On the territory of the Saarland these are the Saarbrücker Hauptsattel and its continuation to the southwest, the Lorraine anticlinorium, northwest of it the Prims syncline and the Nahe syncline as well as southeast the Saargemünder syncline (cf. [111]).

The structural development and filling of the Saar-Nahe Basin is closely associated with a rift phase, which is subdivided into a pre-, syn- and post-phase (cf. [6, 97]). During the pre-rift phase, sedimentation occurred transgressively over the crystalline basement, encountered in the Saar 1 deep well at 5 662 m depth as albite granite, of Middle and Upper Devonian shallow water limestones. These are followed by Lower Carboniferous alum shales. The syn-rift phase, the Permocarboniferous basin development, begins with a basement conglomerate. This is overlain by fluviatile deltaic sediments (sandstones, conglomerates, mudstones) of the Westfal and Stefan with intercalated coal seams (see [99]). Paleomagnetic studies indicate that the Saar-Nahe basin in the Westfal was located immediately north of the equator in the tropical climate range (see [158]). The warm and humid tropical climate in times of delayed or missing subsidence of the basin floor favored the formation of peat bogs, from which later the coal seams developed (see [99]). Not only the special marsh vegetation was important for coal formation, but also an extraordinarily low sedimentation rate (see [157]). The maximum coal flow in the Saar-Nahe Basin has the strata of the Westfal. In the Sulzbach strata (Westfal C), the coal content averages 10% (see [47]). In the Stefan and Unterrotliegend, the number and thickness of coal horizons reduced considerably. The productive sequence of coal is about 4 000 m thick and contains about 130 m of coal thickness that is 3.7 % of the total strata thickness of the mountain. Of this, only about 95 are mineable seams containing 45 m of coal, the thickness of which varies locally by several meters (cf. [151]). At the end of the Stefan, the southeast wing of the Saarbrücker Hauptsattel ruptured as a result of continued constriction in the Saarbrücker Hauptsattel area. This pushed the northwest wing over the southeast wing (cf. [41]) and resulted in a subsidence in the area of the Saargemünder Mulde and the uplift of the Saarbrücker Hauptsattel.

The Rotliegend represents the volcanitic syn-rift phase ($290 - 286$ Ma) and the post-rift phase (thermal subsidence) of the Saar-Nahe Basin and reached thicknesses up to 3 500 m (cf. [98]). The end of the syn-rift phase at the turn of the Lower/Upper Rift is characterized by intense magmatism.

Large-scale basaltic and andesitic lavas formed and numerous rhyolitic-dacitic and andesitic magmas intruded (cf. [97]). The Permocarboniferous deposit exhibits the greatest thickness at the Hunsrück South Rim fault. In the subsequent post-rift phase with thermal subsidence, fine-clastic Rotliegend sediments were accumulated mainly in the area of the Pfälzer and Saargemünder Mulde (cf. [6, 98]). The Rotliegend of the Saar-Nahe Basin is unconformably overlain by the red sandstone, and locally up to 1 900 m of the Saar Basin Rotliegend Formation has been re-eroded (cf. [97, 98]). The Triassic in Saarland (red sandstone and Muschelkalk) belongs to the major distribution of the Germanic facies of Central Europe (cf. [160]). Triassic deposits occur in Lorraine, tongue-like in the Merziger Graben and the Saargemünder Mulde and in the southwest of the Saarbrücker Hauptsattel in the Warndt. While the Saargemünder Mulde shows a broad zone with a clear subsidence tendency in both red sandstone (at least 800 m thickness) and Muschelkalk, the Merziger Graben represents a post-Triassic collapsed part of the Rhenish Slate Mountains (see [189]). Jurassic and Cretaceous deposits are found further west in the Paris Basin. It is assumed that the Jurassic sea might have touched the Saarland, whereas Cretaceous deposits can be excluded (cf. [160]), because in the Cretaceous the Saarland was an ablation area.

In the following Tertiary, too, erosion prevailed in the Saarland. In the Alpidic orogeny, the Mesozoic rock layers were reactivated at the numerous ancient faults. In the Quaternary the present water network developed. Starting from a leveling surface formed in the Oligo-Miocene, rivers and streams cut into the solid bedrock, especially during the ice ages in the Pleistocene. In the process, terraces with sand and gravel deposits formed along the larger flowing waters (Saar, Prims, Blies). Finally, with the formation of loams and slope debris the present morphology was reached (cf. [188]).

On the basis of existing geologic maps and profiles, information on fault zones was summarized in the SPE (PtJ FKZ 0324061) and SYSEXPL projects for the Saarland and presented in the geologic map using the GÜK200 (Geologic Survey Map of the BGR) (see Figure 7.3). The tectonics shown in Figure 7.3 shows, among other things, how strongly especially the Upper Carboniferous, which is exposed on the northwestern flank of the Saarbrücker Hauptsattel, is dissected into floes by numerous faults (cf. [159]). Especially the faults in the southeastern part of the Saarbrücker Hauptsattel are of great geothermal interest because of the high temperatures expected there.

Initial Situation of Geologic Data in Saarland. Usually, for an area such as Saarland, one assumes the availability of a terrestrial "patchwork of measured data" that is not continuous over the area and consists of heterogeneous data types, different data density and accuracy (this situation is common in states

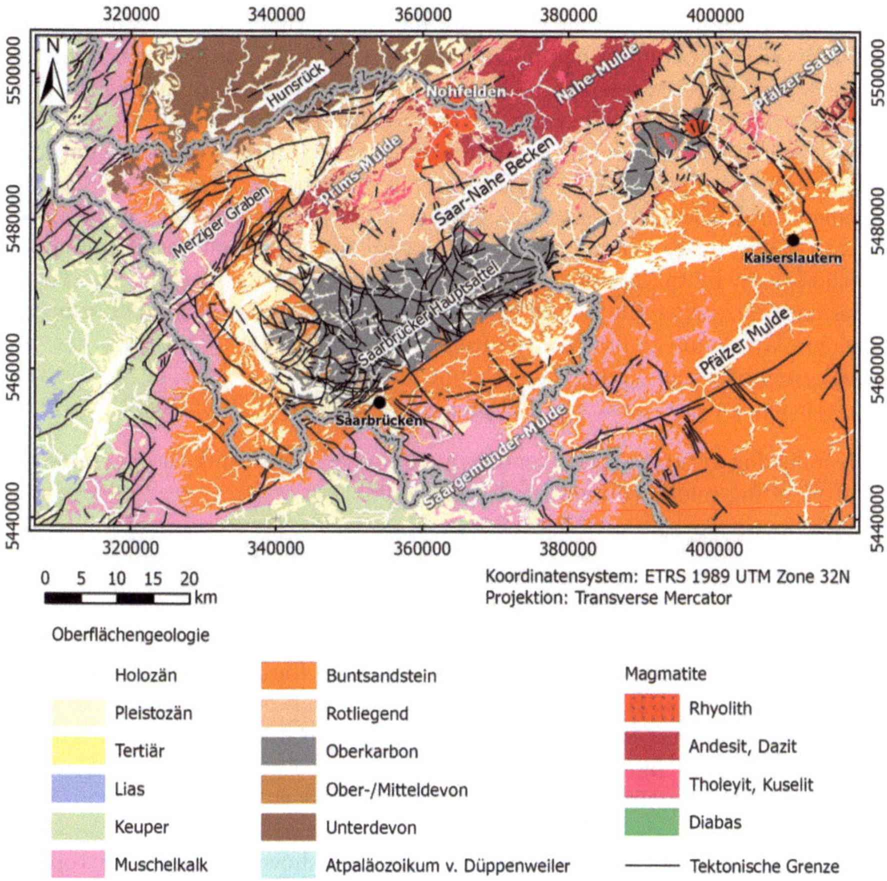

Fig. 7.3 Geologic map with tectonics, modified; data source: GÜK200 - CC 6302 Trier, CC 6310 Frankfurt/Main-West, CC 7102 Saarbrücken, CC 7110 Mannheim, © BGR, Hannover, 2017.

of Germany).

In fact, the data situation in the Saarland region was far more inhomogeneous than expected in respect to distribution and density, date of recording, and accuracy. In order to extract geologically relevant structures for our purposes of decorrelative exploration, the available data must be carefully checked and homogenised, in consistency with the information based on geologic maps. As a consequence, due to the data inhomogeneity of the Saarland area, the preliminary work required a considerable amount of time and effort.

The premise for our desired work is characterized as follows: The quality of the overall result of the data analysis depends largely on how good and complete the available preliminary information is. For this reason, it was first of all advisable to carefully compile all the data with the aim of obtaining as detailed an overview as possible of the geology. Since the assumption of a sufficient amount of usable data for the Saarland unfortunately turned out to be incorrect after careful research, extensive gravimetric measurements of the research group of CBM - Gesellschaft für Consulting, Business und Management mbH, Bexbach, Saarland, became necessary in the end. The purpose was to increase the reliability of a gravimetrically based multiscale potential methodology, by taking into account geologic a priori and new measurement information, so that the risk of success for an appropriate documentation of geologic layers and faults was reduced as far as possible by means of a synopsis, thereby even taking into account the latest findings from alternative exploration sources, such as additionally available seismic information.

The following actions of an inventory were carried out during the PtJ-project SPE (funding reference number: FKZ 0324061):

- *Geologic Maps.* For the presentation of the geologic results, a compilation of preliminary surface information was originally made on the basis of the geologic maps 1:25 000 and 1:50 000 of the Saarland, including their explanations and various hints from former dissertations and publications. On the basis of existing geologic maps and profiles, information on fault zones in the Saarland area was collected and documented in a new geologic map (cf. Figure 7.3) using GÜK200 (Geologic General Map of the BGR) (see Figure 7.3). Furthermore, several dissertations and publications ([41, 97, 159, 160], etc.) were studied for this purpose.

- *Geologic tectonics.* The tectonics in Figure 7.3 shows, among other topics, the extent to which the Upper Carboniferous, which is exposed on the northwestern flank of the Saarbrücken Main Anticline, is broken up into blocks by numerous faults (cf. [159]). Especially, the faults in the southeastern part of the Saarbrücker Main Anticline are of great geothermal interest because of the high temperatures to be expected there.

Magnetics in the Saarland. In the 1960s-70s Dachroth (cf. [36, 37]) carried out numerous geomagnetic works. Literature on the red sandstone area of eastern Saarland he studied, which is adjacent to our study area, was obtained from the State Library team of the Saarland University and State Library in Saarbrücken. From the LIAG (see [181]) the report together with the terrestrial data of the "Ergänzende Magnetikmessungen im Saarland 2008" were provided. Documents on the geology based on the data of the Saar 1 deep borehole were compiled supplementary (see [178, 186]).

Relevant information from magnetic methods in hydrocarbon exploration has been compiled relative to sedimentary rocks (as present in our test areas). Magnetometry along with gravimetry play a primary role in identifying the various subsurface geologic conditions necessary for commercial reservoir development. Both methods are used extensively in coal exploration and to a lesser extent in geothermal resource exploration. The ambiguity of the methods is well-known and understood, but interpretation of their anomalies can be constrained by the use of additional geologic and geophysical data and the joint inversion of data from multiple methods. Ideally, methods based on magnetometry are performed with airborne measurements to cover a large area in the preliminary and regional surveys for geothermal prospecting. Magnetometry in a regional area proves particularly useful in determining the base depth and configuration of sedimentary basins (see [100]).

Significant magnetic anomalies may be caused in hydrocarbon seep environments by diagenetic alteration of magnetic and other iron-bearing minerals. The relationship between sediment magnetic properties and the presence of hydrocarbons is not clear, which may reflect modulating effects of sedimentological, hydrological, and other factors (see [169]). The shape of the anomalies is largely determined by the homogeneity of the anomalous bodies (geologic units). For example, alteration can significantly reduce magnetization because chemical changes also cause magnetic minerals to degrade into less magnetic ones. Inhomogeneity and strong alteration are reflected in large variations of the anomaly (see [177]). Furthermore, the literature and data research particularly addressed the question to what extent the susceptibility of different rock layers can be an additional information to mathematically decorrelate layer boundaries and thus make them useful for a deposit model. For this purpose, the data and literature of the research project FluviMag were evaluated (see [145]). The correlations with magnetics were processed by supplementing them with our own magnetic susceptibility measurements at the Fürstenhausen Ost 1 well. Furthermore, the geomagnetic properties of the deeper geologic layers were elicited by means of susceptibility measurements on existing cores of the Saar 1 deep well.

Mining activities in the Burbach test area. The Oberbergamt of the Saarland provided mine plans from the Luisenthal mine, which is located in the area of the Burbach test area, in digital form for the research work. The individual sheets were scanned and georeferenced, thus the sheets could be scanned with little effort. The provided mine plans contained 48 pit image sheets, which were classified according to the mined seams and levels. From the compiled maps, the study area covering 7.5 km^2 was cut out for further processing. Several borehole profiles were included for processing and validation that are associated with the study area.

There are several coal-bearing strata worth mining in the study area. No mining has taken place in the upper strata such as Anna, Sophie and Cäcilie. In the Burbach area, coal was mined from several seams in the range of -300 mNN to -800 mNN for several decades at the Luisenthal mine through the

Richard shafts. Because the coal-bearing strata in the study area dip from the southeast to the northwest, the mining level in the southeast is closer to the surface (-300 mNN to -400 mNN) than in the northwest area (-600 mNN to -700 mNN).

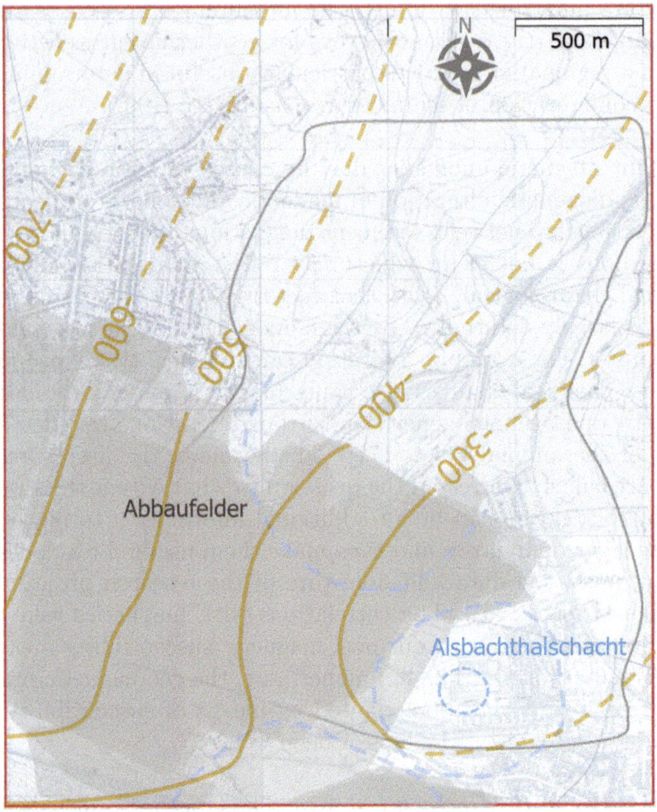

Fig. 7.4 Survey area with mining areas (gray areas), safety pillars (marked with broad blue dashed lines) and isolines for mining height (dashed brown lines). Source: Risswerke der Grube Luisenthal vom Oberbergamt des Saarlandes (see [143]) and Bürgerinformationsdienst (BID) der RAG Aktiengesellschaft (see [147]).

Significant mining activities in the area of the study area took place in the first four of ten seams between 1960 and 1990 (see Figure 7.4). In the southwestern part of the area, mining was particularly intensive in all four coal seams. As a result, the entire southern part of the area is within the mining impact area according to RAG Bürgerinformationsdienst. However, there are no measuring points of the marker leveling, which served for the monitoring of the subsidence of the surfaces. Therefore, the subsidence situation of the surfaces in the study area related to the mining activity is not known. In the northern half of the study area, no significant mining has occurred except in the Ludwig Tunnel, which runs from the east at a depth of -180 mNN to the west at a depth of +100 mNN. In the study area, the southern half is affected by both mining and mining equipment.

Safety pillars were left in place for the stability of the surface installations such as the Alsbach shaft, the Saarbrücken barrage and the Burbach repair company. Nevertheless, mining was partly carried out in the safety pillars, especially in the area of the Burbach repair company (see Figure 7.4). The cavity height created by surface mining initially amounts to 2-3 m, depending on the seam. According to information from the Oberbergamt of the Saarlandes, a backfill mining method was used in the Luisenthal mine.

In this mining method, the cavities are filled with blow backfill immediately after mining operation. This has the advantage that the subsidence of the surface is significantly reduced, depending on the material properties (see [87]). Based on the mining height and the details of the mining areas, the isolines of the coal-bearing seams can be roughly estimated (see Figure 7.4). The information on the solid lines is taken from the mine plans of mining office, and the dashed lines are plausible guesses. The isolines in seam 1 illustrate the distribution and dip of coal seams in the subsurface and are related to conductive horizons beneath the coal seam, which could play an important role in magnetometric interpretation.

Figure 7.5 shows the geology of the Burbach test area.

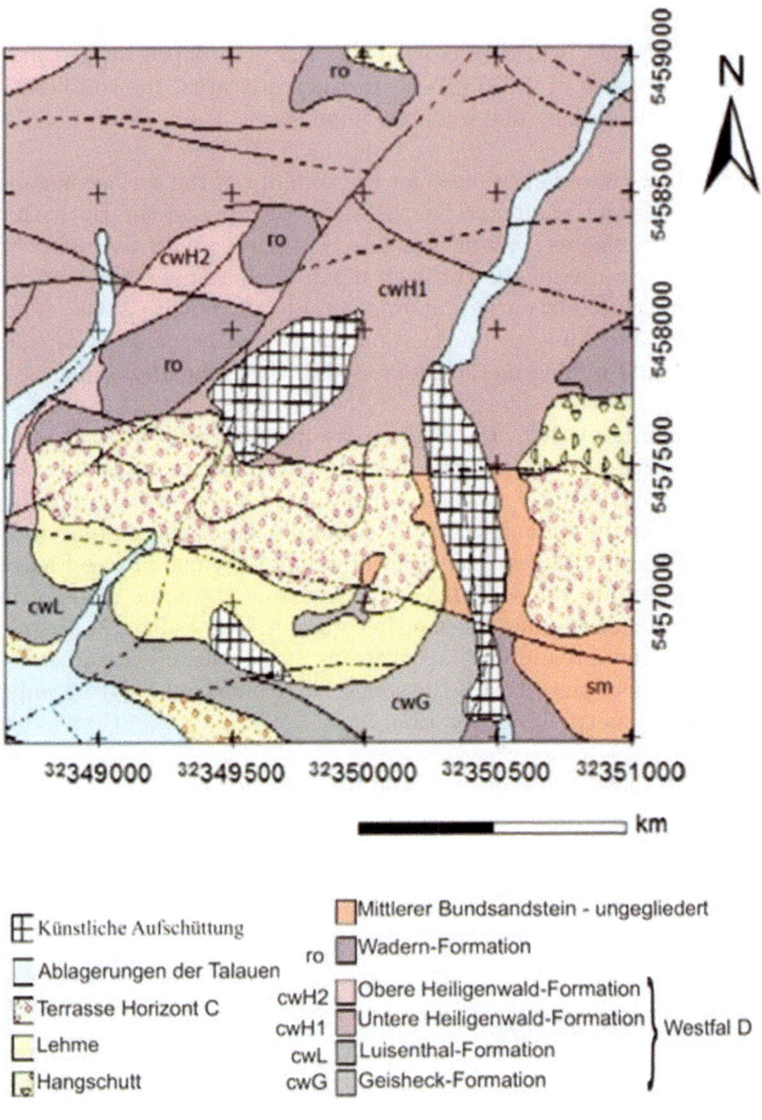

Fig. 7.5 Geology in the Burbach test area.

7.2 Gravimetric and magnetometric data collection

The *gravimetric data collection* describes methods to measure the gravity field of the Earth. In terms of measurement, gravity can be determined, for example, from the period of oscillation of a pendulum or with a torsion balance, although the accuracy is limited to a few millionths of the gravitational force. With modern gravimeters, which work according to the principle of the spring balance, relative measurement uncertainties in the order of $10^{-9}\ ms^{-2}$ are achievable, with superconducting gravimeters even up to $10^{-11}\ ms^{-2}$.

For gravimetric determination of the gravity field, especially for exploration purposes, two different methods are of great interest, namely absolute and relative gravity measurement. For the measurement of absolute gravity, gravimeters operating according to the free fall method are usually used. In this method of measurement, the rate of fall of a test mass is measured with the object (i. e., the "Newtonian apple") moving through a tube in a vacuum. In this manner, an absolute gravimeter determines the absolute value of gravity. A relative gravimeter is a highly sensitive and extremely accurate spring balance: a constant weight stretches a spring according to the gravity prevailing at the measurement location. However, the actual gravity value of a reference point is necessary to determine the gravity. The different linear expansion of the spring is a measure of the corresponding difference in gravity.

In the relative gravimetric method commonly used in exploration today, the amount of gravitational acceleration, i. e., the gravity intensity, is measured. Based on the different values of gravity acceleration at different locations, statements can be made about the distribution of density in the Earth's crust (cf. e. g., [124, 142, 174, 175, 176] for more detailed information).

Gravity measurements can be performed with aircrafts ("airborne measurements"), satellites ("spaceborne measurements"), on the solid Earth surface ("terrestrial measurements"), on the ocean surface ("marine surface measurements"), on the seafloor or (in rare cases) in boreholes. With current satellite technologies it is not possible to investigate finer structures of the Earth's gravity field and Earth's magnetic field, because the signal of a gravimeter/magnetometer attenuates spectrally/exponentially with altitude, which is a direct result of the monopole/dipole integral.

Classical terrestrial *gravity field measurements* are mainly performed in local areas. Very precise geoid calculations or geophysical interpretation of mineral or petroleum deposits or even the detection of tectonic plates can in the foreseeable future only be carried out with terrestrial methods (under certain circumstances also with airborne methods).

The *magnetic survey* is the oldest method of geophysical prospecting, but it has been eclipsed in recent decades by the advent of seismic reflection surveying. Magnetic anomalies can result from a range of changes in lithology, variations in the thickness of magnetized bodies, faults, folds, and topographic relief.

The main objective of magnetometry in geoexploration is to provide information on the relationship between thermal activity, tectonics and stratigraphy of the area by interpreting the magnetic properties in the subsurface. In this context, it should be mentioned that most rocks are not magnetic. However, certain rock types contain enough minerals to cause significant magnetic anomalies. Interpretation of the data to reflect differences in local abundance of magnetization is particularly useful, for example, in locating faults (discontinuities).

Magnetometers for measurement can be roughly divided into two basic types: Scalar magnetometers measure the total strength of the magnetic field to which they are exposed, but not its direction. Vector magnetometers can measure the component of the magnetic field in a particular direction, relative to the spatial orientation of the instrument. Currently, the most commonly used types are the proton recession magnetometer (Overhauser magnetometer), the fluxgate magnetometer, and the optically pumped magnetometer. The proton and fluxgate magnetometers have similar sensitivity (about $1 - 0.1\,nT$), the difference being in the components of the magnetic flux measured. The proton magnetometer measures the total field, that is, the total amplitude of the magnetic field, while the fluxgate magnetometers measure individual components. Fluxgate magnetometers are capable of continuous measurements and are therefore used for measurements from the air, from ships, and from satellites. In contrast, proton magnetometers have no drift and are often used in ground measurements. The optically pumped magnetometers offer much higher sensitivity ($0.001\,nT$) and also high measurement frequency (up to $1000\,Hz$). They are commonly used for archaeological prospection (for more details, the reader is referred to the lecture notes [177] and the references therein).

Magnetic intensity measurements are generally easier to perform than gravimetric intensity measurements, and the measurement procedures do not require a significant amount of time, making it easier to cover large areas.

Exploration method	**Gravimetry**	**Magnetometry**
Type of field	gravity field (monopole field)	magnetic field (dipole field)
Time dependency	time-independent, except for tidal variations	time-dependent
Cause of anomaly field	differences in rock density	differences in magnetization
Characteristics of the maps	gravity map is usually dominated by regional effects	magnetic field map shows a variety of local anomalies
Measuring instruments used	Relative gravimeter (CG-5/-6 Autograv^{TM} from Scintrex Limited)	Overhauser gradiometer/magnetometer (GSM-19 V7.0 from GEM Systems, Inc.)
Measured quantity	gravitational acceleration in $[\mu Gal]$	total intensity in $[nT]$ gradient of total intensity in $[nT/m]$
Measuring method	passive measuring method	passive measuring method
Resolution	1 μGal (CG-5), 0.1 μGal (CG-6)	0.01 nT (magnetic field and gradient)
Accuracy	< 5 μGal (CG-5/CG-6)	< 0.1 nT
Weight	8 kg (CG-5 incl. battery) 5.5 kg (CG-6 incl. batteries)	2.1 kg (console), 1.1 kg (sensor), 0.9 kg (rod)
Personnel effort	2 persons (height measurement) 2 persons (gravity measurement)	2 persons
Estimated time required (based on measurements from SPE and SYSEXPL)	measurement point spacing $10 - 20$ m 60 - 90 measuring points/10-hour working day	measurement point spacing 10 m 600 - 900 measuring points/10-hour working day

Table 7.1 Rough tabular comparison of gravimetry and magnetometry with respect to the instruments used in SPE and SYSEXPL.

For a test of the novel mollifier methods by means of high-precision measurements in selected test areas, the company CBM - Gesellschaft für Consulting, Business und Management mbH, Bexbach, Saarland, – as already in the preliminary SPE project (PtJ FKZ 0324061) for gravimetry – had to perform own measurement campaigns. For the cleanup of the measured data, the usual and well-known correction methods in gravimetry and magnetometry were used (cf. PtJ final report on SPE (PtJ FKZ 0324061) for the case of gravimetry). Table 7.1 shows a comparison of the measurement methods in gravimetry and magnetometry (cf. also [82, 109, 161, 162, 170]).

7.3 Gravimetric measurement campaigns

The aim of the measurement campaigns in the test areas was the acquisition of potential data in a narrow grid, which allows an inversion into the depth. Within the scope of the predecessor project SPE (PtJ FKZ 0324061), a gravimetric measurement campaign of CBM took place in the test area Burbach (Saarbrücken). Gravimetric measurements require a high degree of care, especially when determining the height of the grid points, since their accuracy has a great influence on further calculations. The CBM's own measurement campaign consisted of a total of four parts, the sequence of which is shown in Figure 7.6.

The measurement point grid to be processed was set at 50 m in the CBM projects, taking into account topography and forestation. The grid points were staked out predominantly in forested areas and were carried out by means of traditional tacheometry. In this way, it was possible to ensure compliance with the grid (see Figure 7.7).

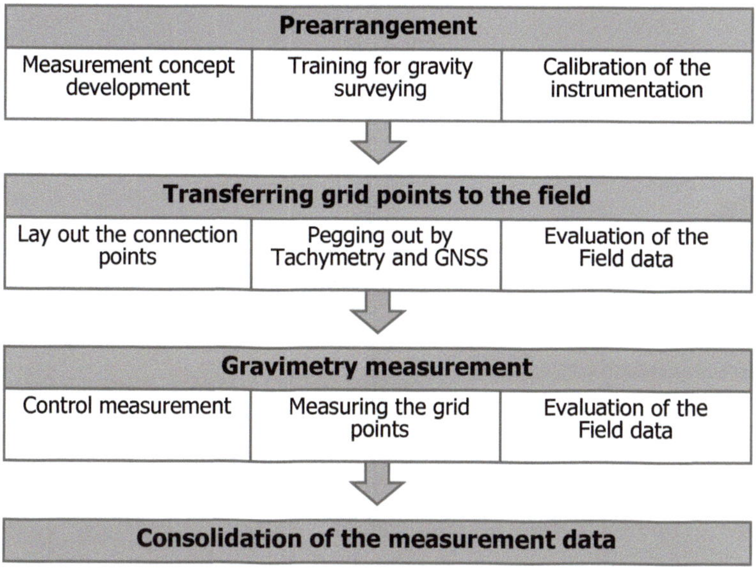

Fig. 7.6 Methodical procedure of the gravimetric measurement campaign.

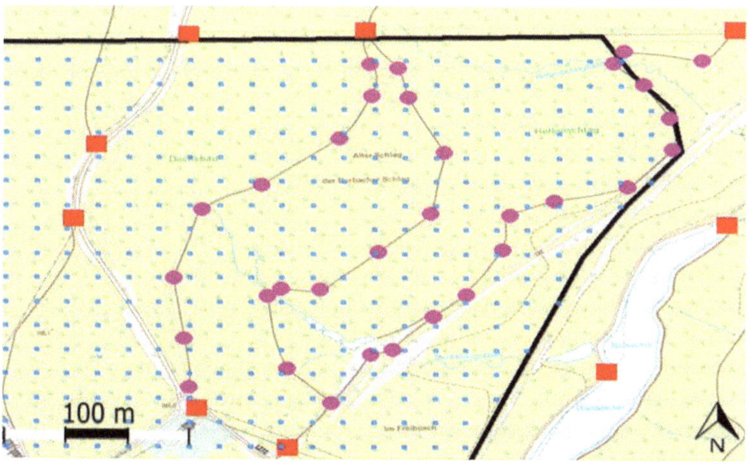

Fig. 7.7 Map section with points: GNSS fixed points (red), polygon points (purple), grid points in 50 m distance (blue).

Fig. 7.8 Relative gravimeter CG-6 in use (CBM-measurement campagne 2017, scientifically supported by the State Office for Surveying and Basic Geographic Information Rhineland-Palatinate, Koblenz).

Two Scintrex Limited relative gravimeters were used to determine the gravity values, a CG-6 Autograv^{TM} (see Figure 7.8) and temporarily a CG-5 Autograv^{TM}. The inaccuracy (simple standard deviation) in the determination of the absolute gravity of the new measurements was about 27 μGal. This value was obtained as the sum of the following components: Accuracy of the connection point in the Saarland gravity network (10 μGal), measurement accuracy of CG-5 and CG-6 (without repeat measurements: 7 μGal), uncertainty in the open air reduction from sensor height to ground surface (10 μGal). Instrument height and northing were determined using a tape measure and compass. At the beginning of the measurement campaign, the Landesvermessungsamt Baden-Württemberg ensured at several prominent measurement points that the new measurements were recorded and processed in a fully compatible way with the existing ones of the Landesamt für Vermessung, Geoinformation und Landentwicklung (LVGL), Saarland.

Gravimetric adjustment methods were further developed by Dr. T. Degro on the basis of preliminary work by LIAG. More details can be found in the final report of the research project SPE (PtJ FKZ 0324061).

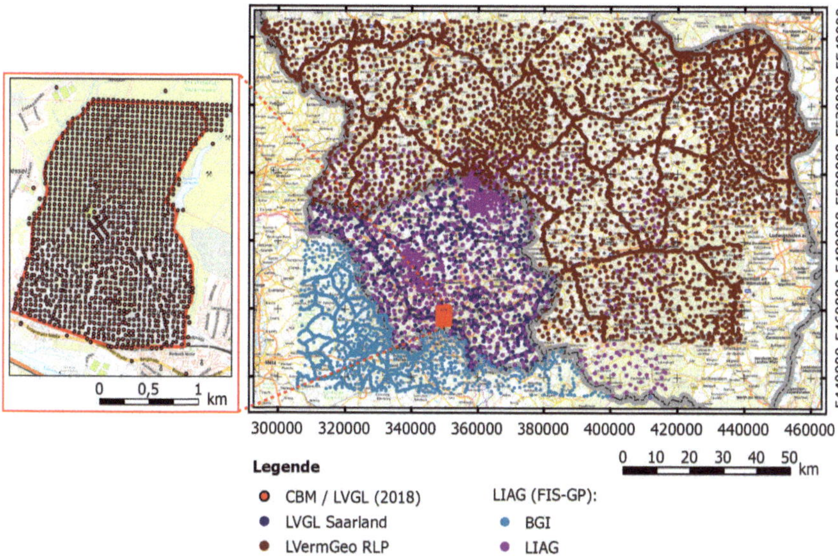

Fig. 7.9 Overview map of all data points that were available for decorrelation. The enlarged red box shows the map section of the measurement performed by CBM 2018; data basis: "Data source: gravity measurements, specialized information system geophysics, or anomalies of the total geomagnetic field. © Leibniz Institute for Applied Geophysics (LIAG), Hannover, 2017." "Geobasisdaten, © LVGL V 2302 B - SPE" - Copyright: Landesamt für Vermessung, Geoinformation und Landentwicklung Saarland, * Source: State Office for Surveying and Geobasis Information Rhineland-Palatinate (12/2019), Map basis: © GeoBasis-DE / BKG (2019).

During the measurement campaign, additional points from the LIAG dataset were integrated and additional selected connecting lines were laid from the measurement area to the already measured surrounding areas, thus ensuring a seamless transition between the individual measurement campaigns and to minimize possible oscillations at the measurement area transitions. It is important to note that local artificial high-frequency isoline shearing has a negative effect on the quality of the model inversions.

During the CBM measurement campaign, 1 502 data points were recorded. A total of 11 916 data points were available for the gravimetric surveys after the completion of data acquisition, which are color-coded by data origin in the map in Figure 7.9 (see also Table 7.2).

Data origin	Data points
LVermGeo RLP	4 681
LVGL Saarland	1 201
FIS-GP	4 532
• LIAG	
• BGI	
CBM/LVGL (2018)	1 502
Total	11 916

Table 7.2 Listing of the used data points according to the data origin.

In order to ensure uniqueness of a geologic structure, additional preliminary information,e. g. on geology, must be used, which is available especially in the Saarland from borehole information from the time of mining. Coal mining and the iron industry were the most important economic sectors in the Saarland. The formation of the Saarland and its changeable political history are based on the geologic conditions of its subsurface (a more detailed description of the geology in the Saarland can be found in section 7.1).

7.4 Magnetometric measurement campaigns

To carry out the magnetometric measurement campaigns, three types of measuring instruments were used in the project:

- GNSS measuring device for position determination,
- Overhauser magnetometer for measuring total intensity,
- 3-component gradiometer for recording magnetic field strengths in three orthogonal directions.

The magnetometric data collection was carried out in several measurement campaigns with the Overhauser magnetometer in the test areas Burbach and Nohfelden. The methodical procedure of the magnetometric measurement campaign is shown in Figure 7.10.

With the data acquisition at the Earth's surface by means of a 3-component gradiometer, new scientific territory was entered in SYSEXPL. The measurement apparatus had to be adapted several times and data cleaning also proved to be a major challenge. Nevertheless, in order to ensure uniqueness of a geologic structure, additional prior information, e. g. information on susceptibility, had to be obtained. For this reason, susceptibility measurements were performed for the evaluation in the Burbach measurement area (see section 7.7).

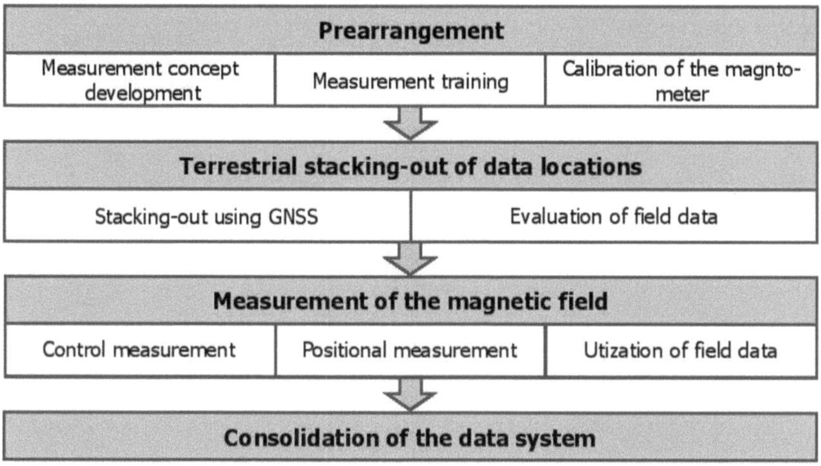

Fig. 7.10 Methodical procedure of the magnetometric measurement campaign.

Magnetometric measurement campaigns using Overhauser magnetometer. In the Saarland, measurements with the Overhauser magnetometer were performed in two areas (see Figure 7.11), one of them near Eisen, a district of the municipality of Nohfelden, and the other one in Burbach, a district of Saarbrücken.

7.5 Measurement campaign Nohfelden

Measurements in the Eisen district of Nohfelden took place from Oct. 18 to Oct. 25, 2019, and were extended in sequence on Feb. 18, 2020, to include the two profiles located south of the measurement area (see Figure 7.12).

The two Overhauser magnetometers and the GNSS system were used in the Nohfelden measurement campaign. The aim was to perform magnetic measurements at the positions of the gravimetric measurements, which were known from the SPE project (PtJ FKZ 0324061) and were provided by the Leibniz-Institut für Angewandte Geophysik. For this purpose, the coordinates of the gravimetric measurements (see Figure 7.12, left) were remeasured with the GNSS system and then the total intensity and gradient were recorded with the GSM-19.

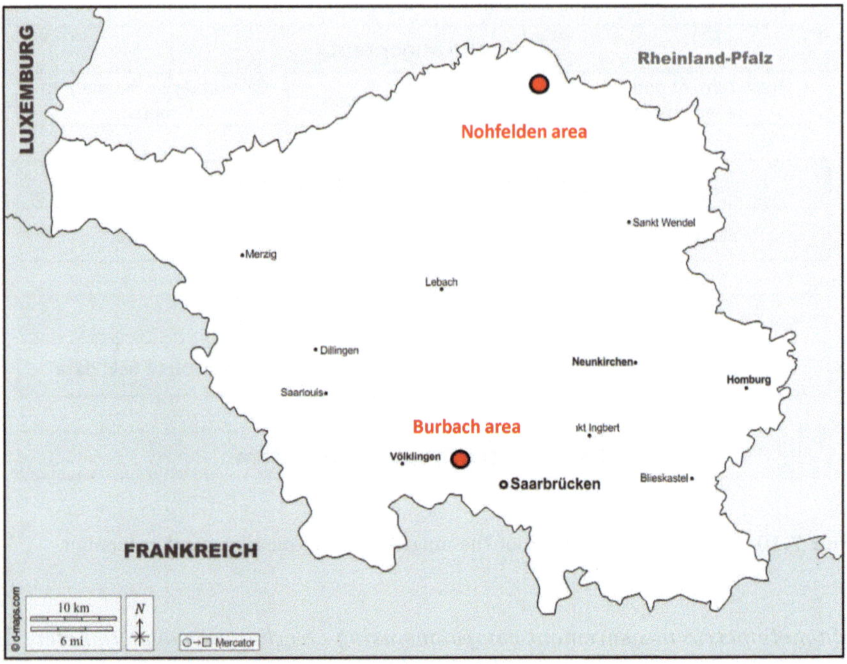

Fig. 7.11 Overview map of the measurement areas (red dots) located in the Saarland. Of these, one is located in the municipality of Nohfelden and the other in Burbach, a district of Saarbrücken. Map basis: © d-maps.com.

Additional measurements were performed to increase the point density. The base station had to be set up at three different locations (see red crosses in Figure 7.12, right) due to anthropogenic influence during the campaign.

7.6 Measurement campaign Burbach

The measurement campaign in Burbach, a district of Saarbrücken, was carried out in a total of four sections in the period from March 2020 to May 2021 with the aim of a synoptic observation based on the gravimetric measurements from the SPE project (PtJ FKZ 0324061) and the magnetometric measurements from the SYSEXPL project (see Figure 7.13).

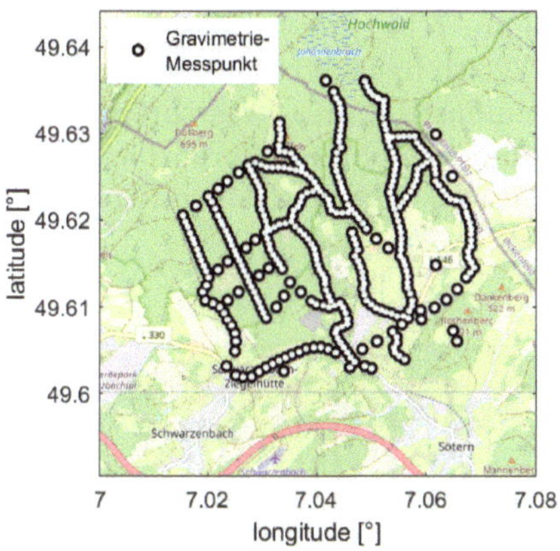

Fig. 7.12 General map of the measurement area near Eisen, a district of the municipality of Nohfelden, with the measurement positions of gravimetry (left) and magnetics (right), as well as the positions of the magnetic base station (right, red crosses). Map basis: © OpenStreetMap contributors.

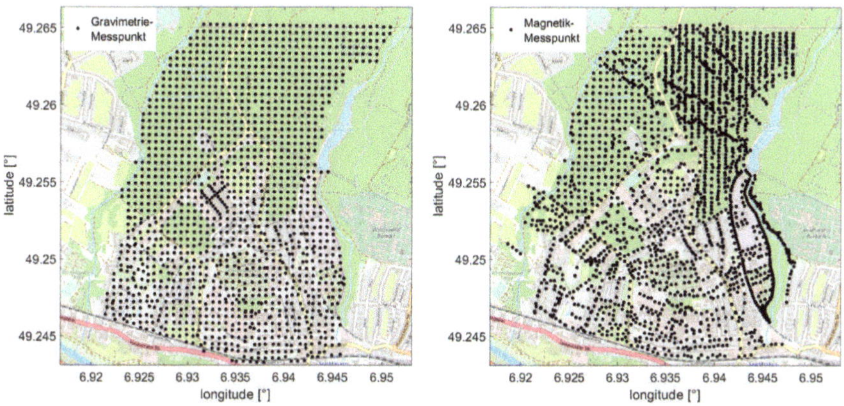

Fig. 7.13 General map of the measurement area in Burbach, a district of Saarbrücken, with the measurement positions of gravimetry (left) and magnetics (right). Map basis: © OpenStreetMap contributors.

The first Overhauser measurements took place from 03/03/2020 to 03/13/2020 and were carried out in cooperation with the LVGL Saarland in the northeastern measurement area, where the 50 m x 50 m grid used in gravimetry was reduced to 25 m in the N-S direction (see Figure 7.14 (black)). In addition, three profiles were placed over the Saar fault and four profiles over the East fault to obtain more detailed information regarding contrast and resolution. Similarly, the probability of success of a measurement in the development area should be estimated. The 2nd measurement section ran from 05/12/2020 to 06/10/2020 and had the goal to complete the remaining, undeveloped area in the northwest (see Figure 7.14 (gray)). Subsequently, in the 3rd measurement phase from 03/17/2021 to 04/01/2021, the built-up area located in the south, including the industrial area, was surveyed (see Figure 7.14 (blue)). During a final measurement phase from 05/03/2021 to 05/07/2021, gaps in the area of the 3rd section were closed (see Figure 7.14 (red)).

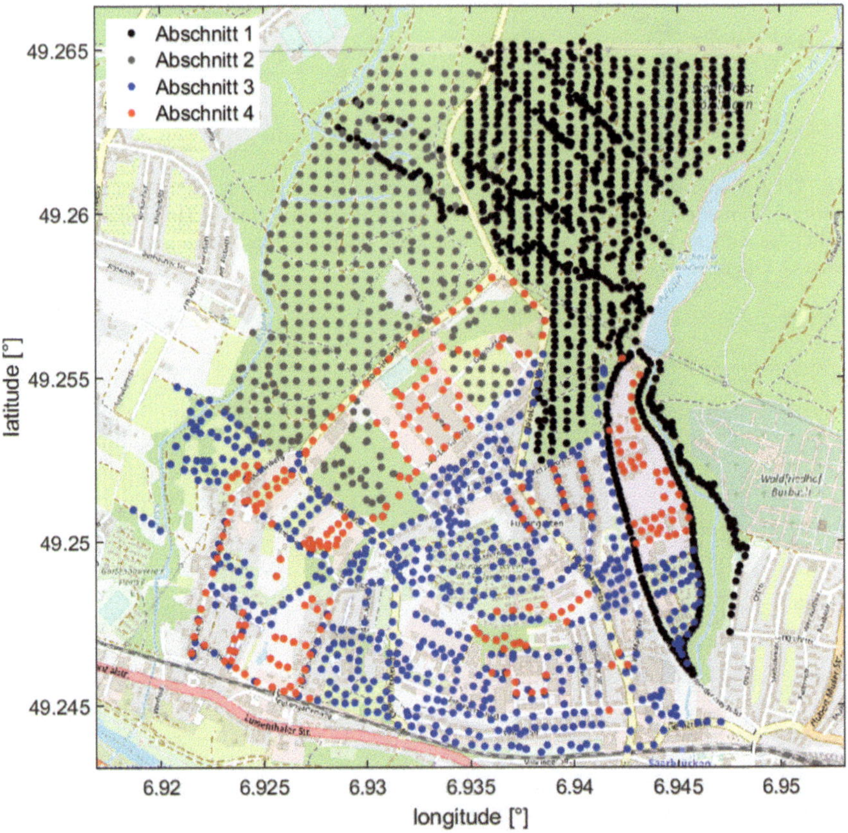

Fig. 7.14 Subdivision of the measurement area in Burbach into the four sections in which they were measured in time: 03/03.-03/13/2020 (black), 05/12/-06/10/2020 (gray), 03/17-/04/01/2021 (blue) and 05/03/-05/07/2021 (red). Map basis: © OpenStreetMap contributors.

7.7 Susceptibility measurements as pre-information

Fig. 7.15 Location of surveyed samples as part of susceptibility measurements in September 2021. Map basis: © Google Earth.

No uniqueness of an interior geologic structure can be obtained from only terrestrial magnetometric data. Therefore, for the mollifier multiscale methods used to represent the geology in section 6.3, additional preliminary information inide the study area is required. This supplementary information includes magnetic susceptibility, which is linearly incorporated into the magnetization. Therefore, measurements of the magnetic susceptibility in the Burbach measuring area were carried out with an MS2K from the manufacturer Bartington. The MS2K instrument is a "handheld sensor" that guarantees well reproducible measurements of the magnetic volume susceptibility for reasonably smooth sample surfaces. After initial measurements on samples from the Bexbach Mining Museum, observations were made on the surface in the Burbach measurement area (see Figure 7.15) and further measurements were made at a suitable location south of the measurement area.

Since measurement samples in the field are naturally affected by weatering and could only be obtained from the surface, samples from the Zentrum für Biodokumentation of the Saarland in Reden were additionally examined. In the process, rock samples from various depths from the mines in St. Ingbert, Luisenthal, Jägersfreude, St. Barbara, Dechen, Velsen, Kohlwald, Helene, Göttelborn, Merlenbach, Heinitz and Maybach as well as samples from Riegelsberg and Leitersweiler could be measured, with a focus on rock types that are also expected in the Burbach subsurface.

Finally, samples of the drill core stock in Jägersfreude of the RAG Aktiengesellschaft could be examined, whereby the drill cores of the Fürstenhausen Ost well were measured due to their proximity to the measurement area. The Fürstenhausen Ost well was drilled in 1972, has a depth of 1355.60 m and extends over the Heiligenwalder strata, Luisenthaler strata, Geisheck strata, Sulzbacher strata and extends into the Rothell strata. The relevant drill samples were measured with the MS2K and information on all mentioned layers could be obtained. In addition, information on deeper layers could be obtained by surveying samples from the Saar 1 deep well.

Chapter 8

Geologic mollifier decorrelation applied to Saarland test areas

Abstract:

The aim of decorrelative gravimetry and/or magnetometry is to convey the basic understanding that zooming-in potential methods using mollifier potential wavelets open up new important fields of application in today's geoexploration, especially for areas with mining-related cavities or very dense development, such as the Saarland, which make the use of reflection seismic measurements difficult or even impossible. Decorrelative gravimetry as well as magnetometry, each in its own right as well as in combination, also provide an initial assessment of possible prospects in regions where little other, e. g., seismic, data are available.

In this chapter, the mollifier decorrelation methods are applied, exemplary for anthropogenically influenced areas, to a test area of the Saarland near the Saarbrücken district of Burbach. We provide an overview of the main results by geomathematically analyzing methodologies for the geologic structure on the surface as well as in depth.

W. Freeden, H. Nutz, *Exploratory Potential Methods in Geothermal Power Generation*,
Lecture Notes in Geosystems Mathematics and Computing,
https://doi.org/10.1007/978-3-031-54412-5_8

8.1 Geologic mollifier decorrelation using surface methodologies

Based on gravimetric and magnetometric measurements (see also Table 7.1), a comparative analysis using the developed decorrelation methods is now performed for the following for three areas, viz.

- Greater Rhineland-Palatinate/Saarland,
- test area Nohfelden (Saarland),
- test area Burbach (Saarland).

Surface interpretation of the greater Rhineland-Palatinate/Saarland. The developed decorrelation techniques were first performed on the total geomagnetic field and compared with the gravitational anomaly decorrelation methods known from SPE (PtJ FKZ 0324061).

Figures 8.1 and 8.2 represent the gravitational (Bouguer) and magnetic field anomalies in the greater Rhineland-Palatinate/Saarland. The corresponding representation of the geology in this area can be found in Figure 7.3.

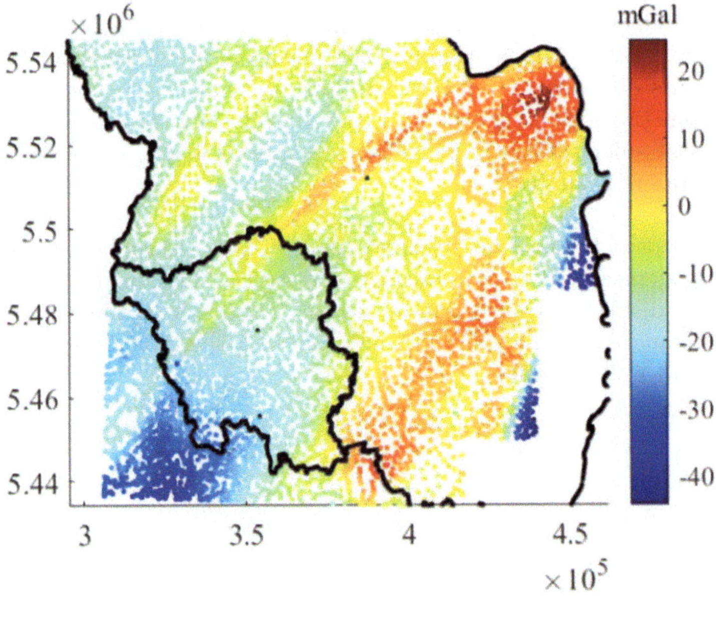

Fig. 8.1 Measured values of gravitational anomalies in the greater Rhineland-Palatinate/Saarland for 11 899 data points.

A dataset of anomaly values of the total geomagnetic field (in nT) in 1 795 246 data points is available in the greater Rhineland-Palatinate/Saarland (cf. Figure 8.2). This dataset was provided by the Leibniz Institute for Applied Geophysics (LIAG) in a 100 m grid. These refer to the DGRF 1980 geomagnetic reference field, epoch 1980.0, at a height of 1 000 m. Unfortunately, this dataset does not consist of actual measurements but of interpolated values based on an airborne campaign.

The decorrelation of the anomalies of the total geomagnetic field is a great challenge. In addition to smaller magnetic anomalies in the Nohfelden area, much larger anomalies are detectable in the eastern part of the Palatinate Forest. The most pronounced similarity of gravity anomalies with the magnetic field is reflected in the Hunsrück-Südrand fault. This is characterized as a SW-NE-trending magnetic band structure with superimposed low-amplitude short-wavelength anomalies, and it stands out as a sharper, discontinuous SW-NE lineament in the bandpass filtering of the gravity anomalies (see Figures 8.3 to 8.5).

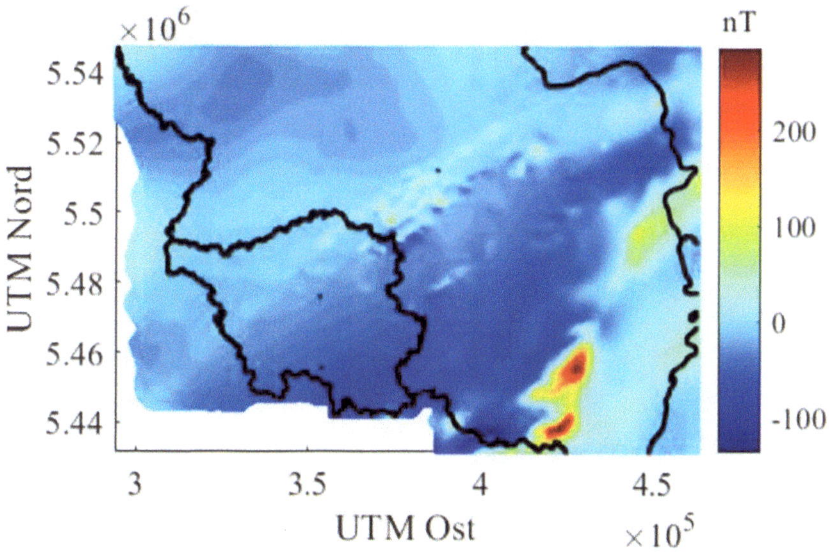

Fig. 8.2 Magnetic field anomalies in the greater Rhineland-Palatinate/Saarland based on 1 795 246 data points.

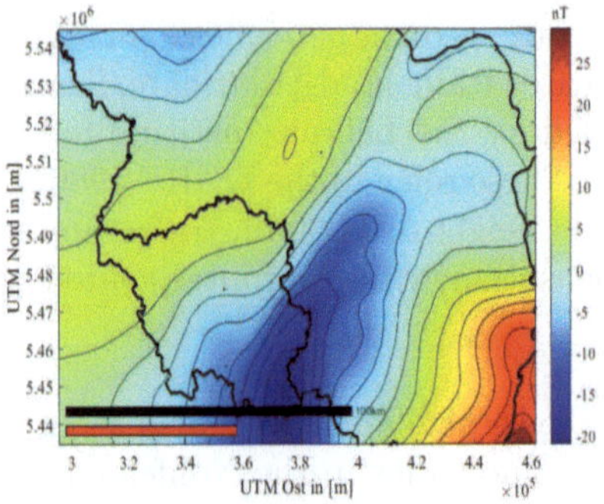

Magnetic anomaly scale $j = 4$

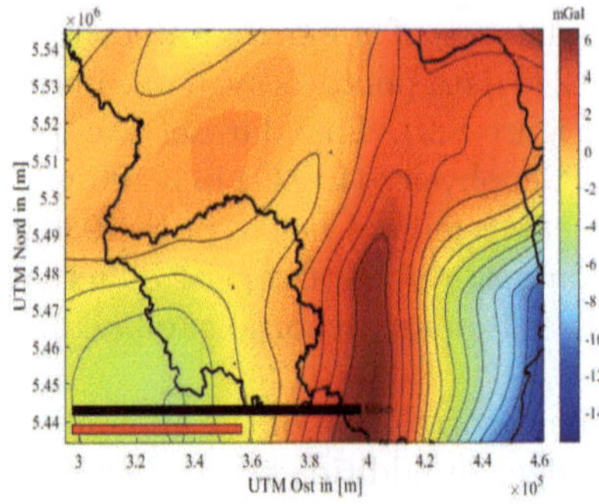

Gravitational anomaly scale $j = 4$

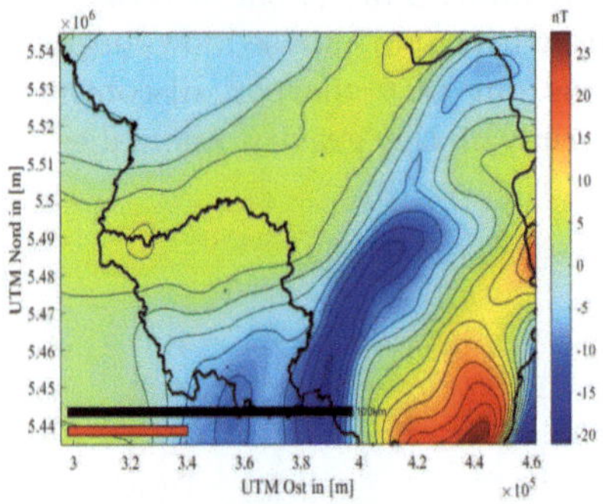

Magnetic anomaly scale $j = 5$

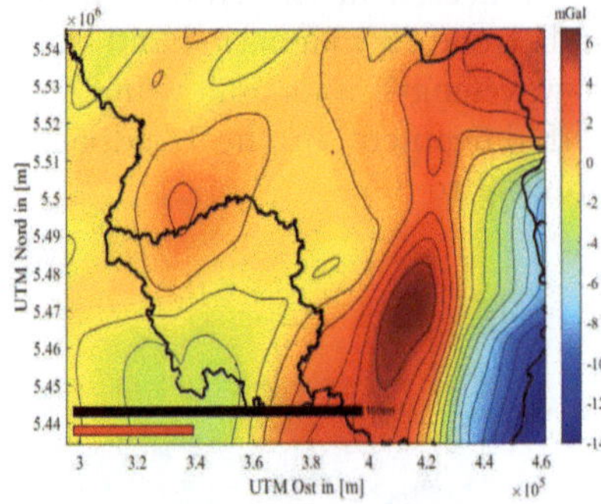

Gravitational anomaly scale $j = 5$

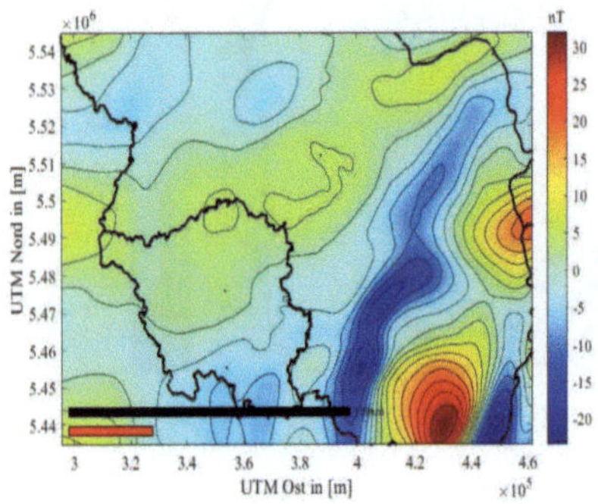

Magnetic anomaly scale $j = 6$

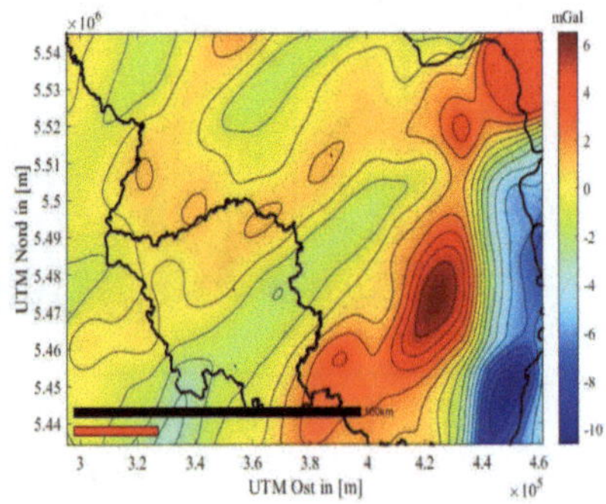

Gravitational anomaly scale $j = 6$

Fig. 8.3 Comparison of the bandpass-filtered magnetic field anomaly (left) and the bandpass-filtered gravitational field anomaly (right) in the greater Rhineland-Palatinate/Saarland.

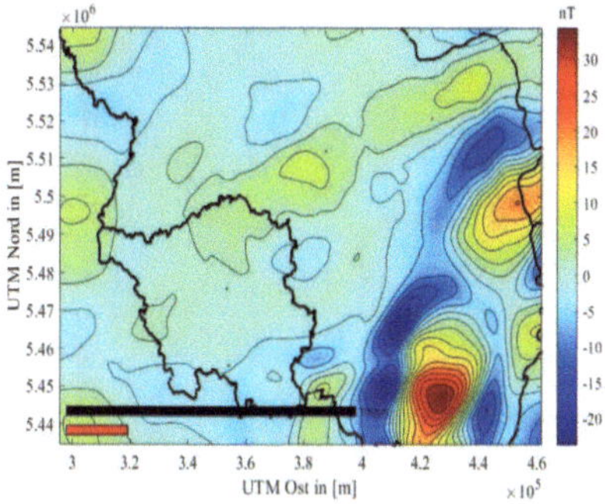

Magnetic anomaly scale $j = 7$

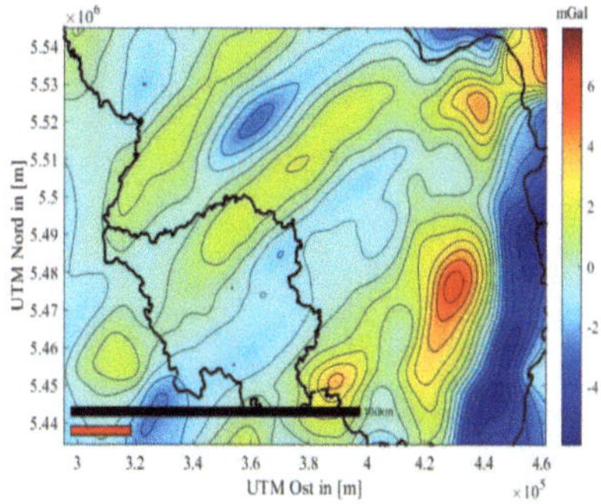

Gravitational anomaly scale $j = 7$

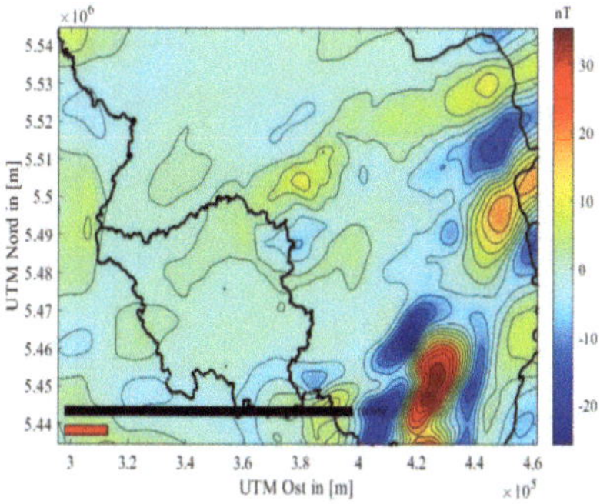

Magnetic anomaly scale $j = 8$

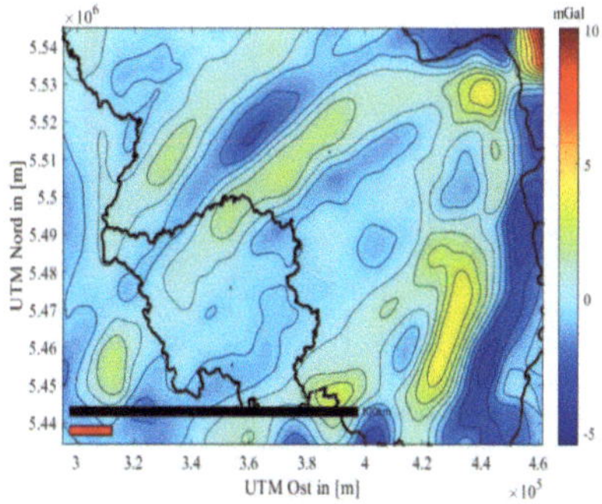

Gravitational anomaly scale $j = 8$

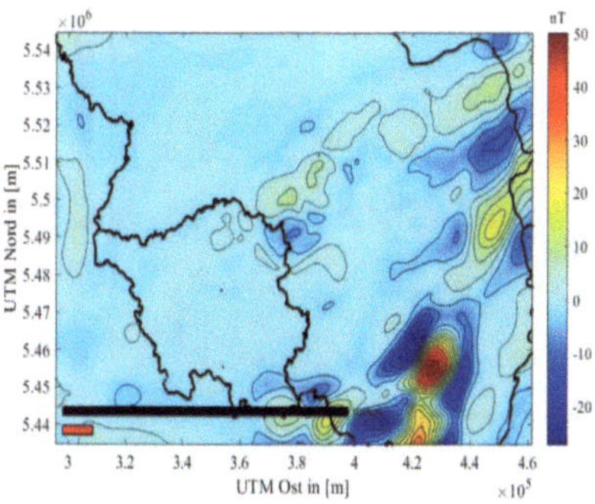

Magnetic anomaly scale $j = 9$

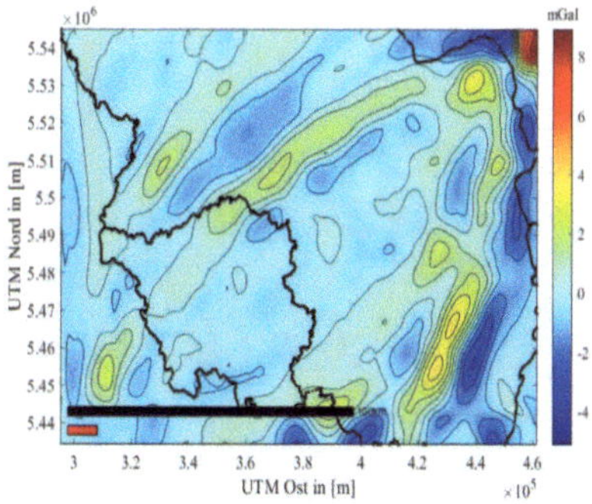

Gravitational anomaly scale $j = 9$

Fig. 8.4 Comparison of the bandpass-filtered magnetic field anomaly (left) and the bandpass-filtered gravitational field anomaly (right) in the greater Rhineland-Palatinate/Saarland, continued.

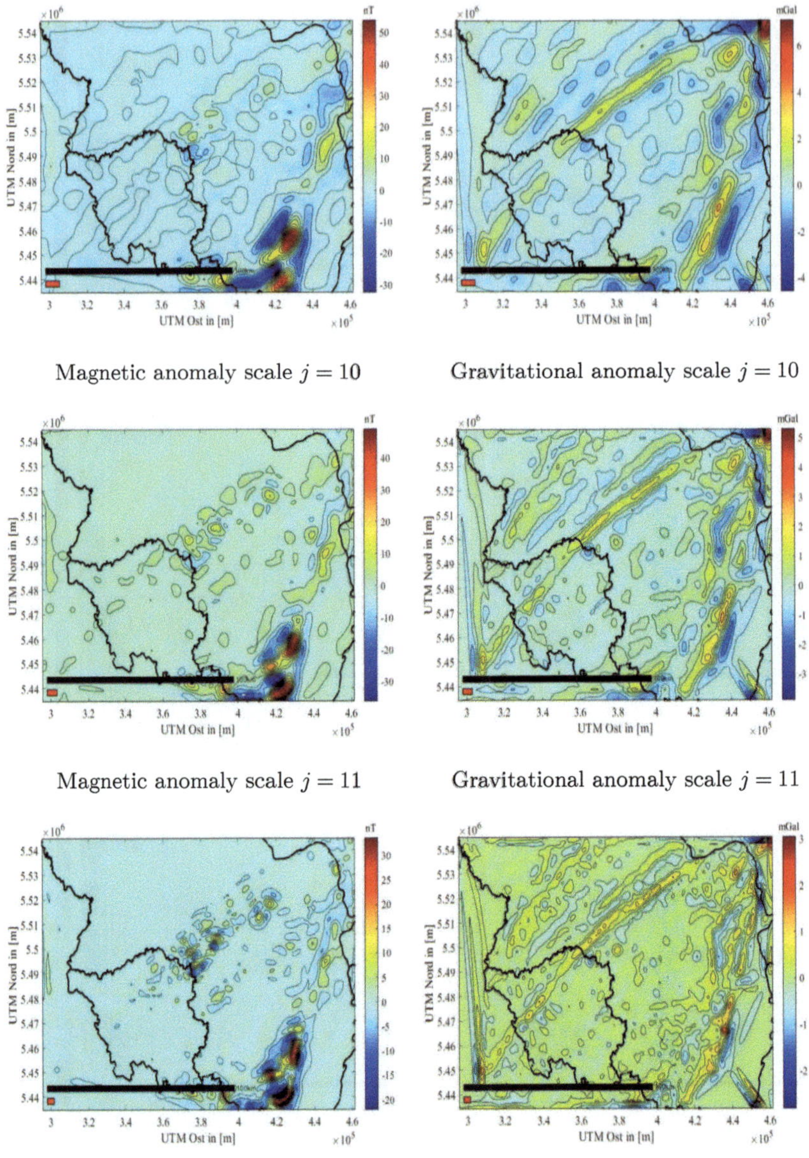

Magnetic anomaly scale $j = 10$ Gravitational anomaly scale $j = 10$

Magnetic anomaly scale $j = 11$ Gravitational anomaly scale $j = 11$

Magnetic anomaly scale $j = 12$ Gravitational anomaly scale $j = 12$

Fig. 8.5 Comparison of the bandpass-filtered magnetic field anomaly (left) and the bandpass-filtered gravitational field anomaly (right) in the greater Rhineland-Palatinate/Saarland, continued.

Surface interpretation of the Saarland test area Nohfelden. In the Nohfelden area a dataset of magnetic field anomalies consisting of 1 808 data points is available (cf. Figure 8.7). One of the reasons for the selection of this area is the availability of a corresponding dataset concerning gravity data consisting of 304 data points (cf. Figure 8.6). This comes from the data pool of SPE (PtJ FKZ 0324061) and is a part of the overall gravimetric dataset. A direct comparison between magnetic field anomalies and gravitational field anomalies is realizable and practicable.

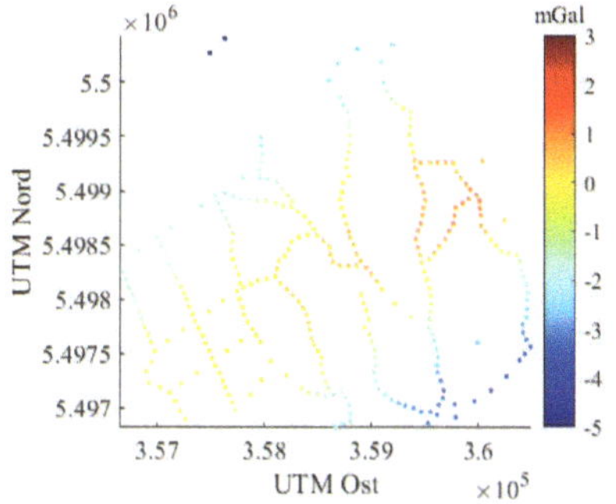

Fig. 8.6 Measured values of gravitational field anomalies in the Nohfelden area for 304 measurement points.

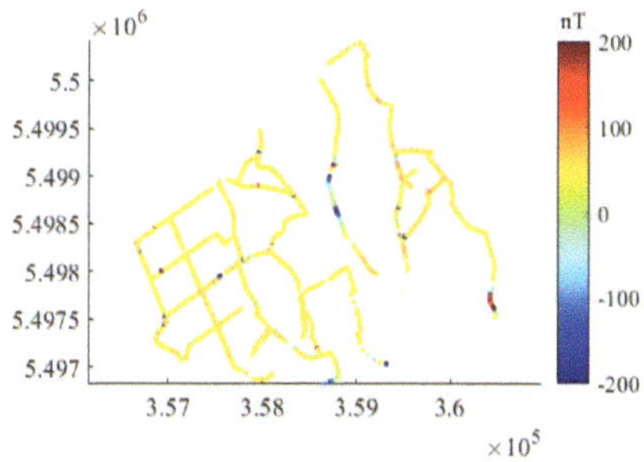

Fig. 8.7 Measured values of magnetic field anomalies in the Nohfelden area for 1 808 measurement points.

The same mollifier methodology as in the greater Saarland/Rhineland-Palatinate was applied to the Nohfelden area, where the results were interpreted using the geologic map in Figure 8.8 (see Figures 8.9 to 8.11). There, the SW-NE strike direction of the geologic layers in the Nohfelden area is visible from scale $j = 4$ to $j = 8$ in the decorrelation of the gravity anomalies. The elongate positive anomaly is separated from the negative elongate anomaly by the Hunsrück-Südrand fault. At scale $j = 7$ the slope debris is clarified. At $j = 8$, the pointed Taunus quartzite with negative anomaly and fluviatile valley deposits can additionally be identified. At scale $j = 4$ of the decorrelations of the magnetic anomaly, the northwest part can be differentiated from the southeast, so that the Hunsrück shale in the northwest is to be considered together with the slope debris. The large positive anomaly indicates the andesite. At scale $j = 5$ the figure is shaped by the influence of the slope debris. This is enhanced at scale $j = 6$ and in the center of the figure the Hunsrück shale clearly emerges with negative anomaly. In the northwest, at scale $j = 7$, the different geologic layers can be recognized: Quaternary and Lower Rotliegend. In the southeast, it is the Andesites. These can be better localized at $j = 8$ and $j = 9$. Likewise, the Lebach strata are better separable from the Kusel strata in some areas due to the irregularly occurring spheroidal siderites. For the interpretation of the southeastern area and thus for the identification of the large fault a condensation of the data is necessary. The clearly positive gravity anomaly in the area of the Kusel formation is enhanced by the dipole effect of the magnetic field anomaly.

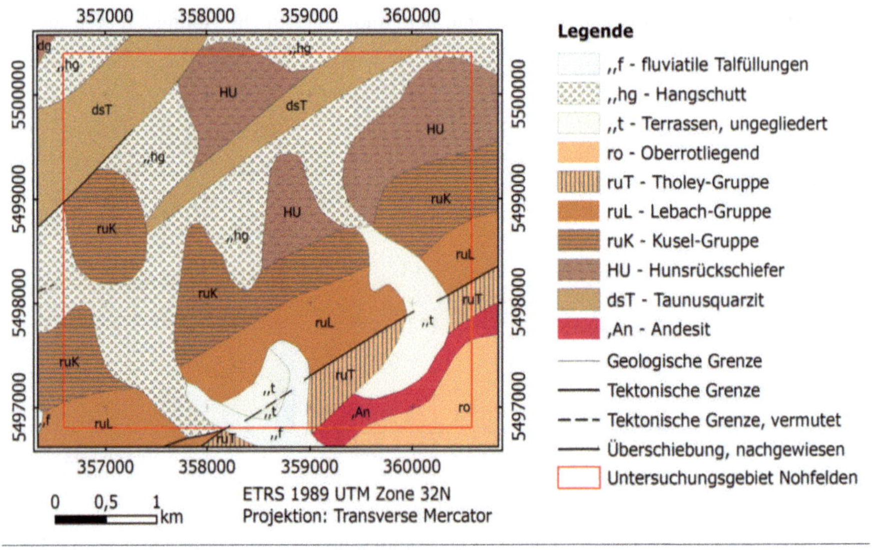

Fig. 8.8 Surface geology in the Nohfelden area (data source: GüK200 - CC6302 Trier, c BGR, Hannover, 2017).

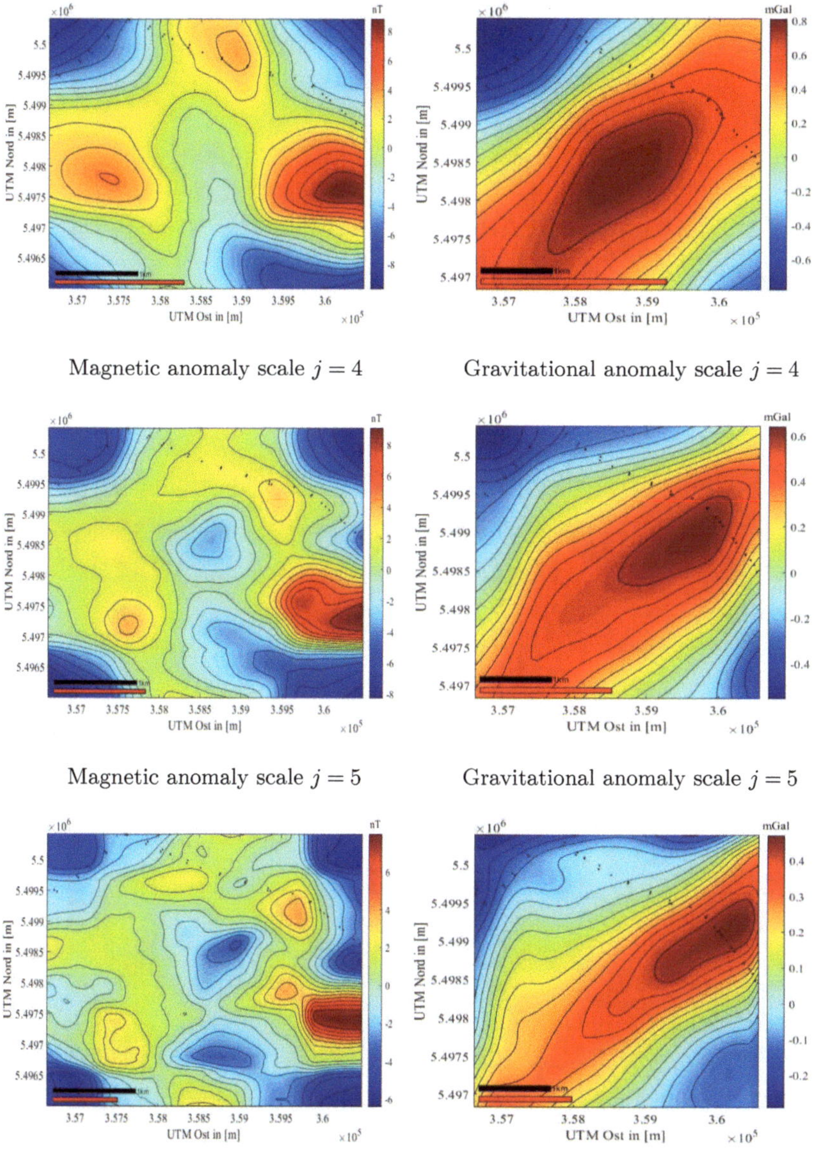

Magnetic anomaly scale $j = 4$ Gravitational anomaly scale $j = 4$

Magnetic anomaly scale $j = 5$ Gravitational anomaly scale $j = 5$

Magnetic anomaly scale $j = 6$ Gravitational anomaly scale $j = 6$

Fig. 8.9 Comparison of the magnetic field anomaly (left) and the gravitational field anomaly (right) in the Nohfelden test area.

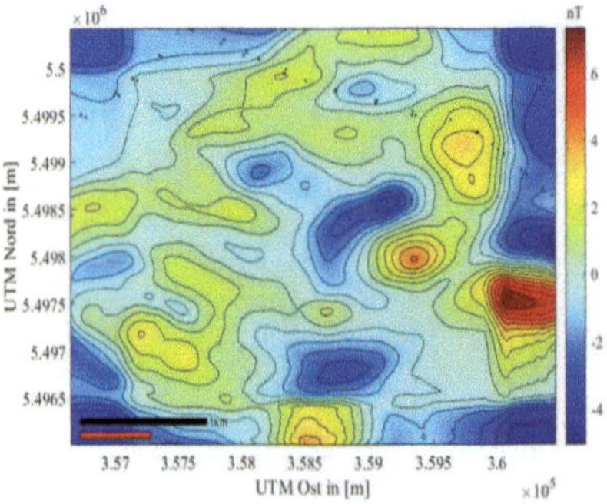

Magnetic anomaly scale $j = 7$

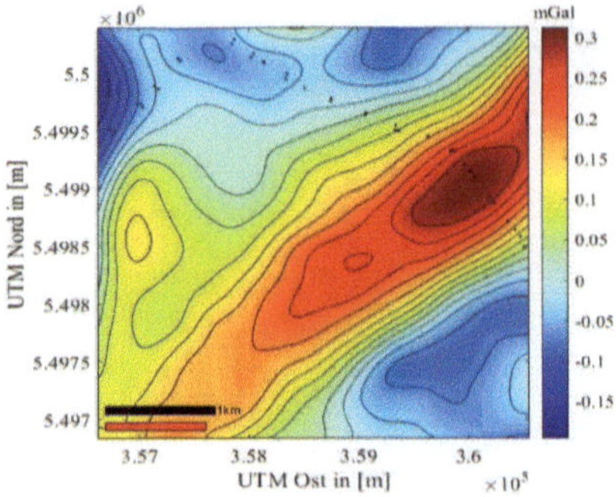

Gravitational anomaly scale $j = 7$

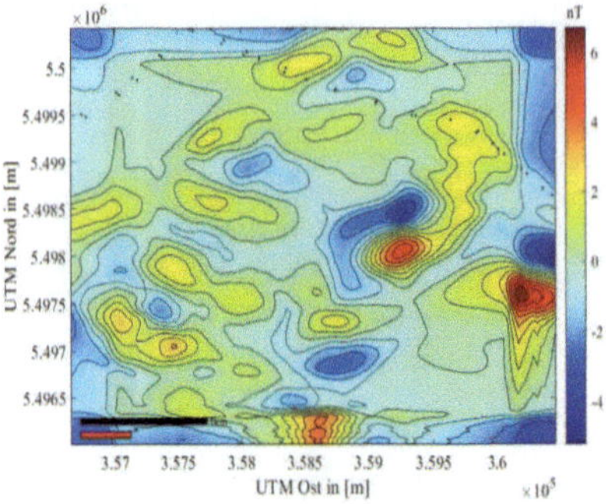

Magnetic anomaly scale $j = 8$

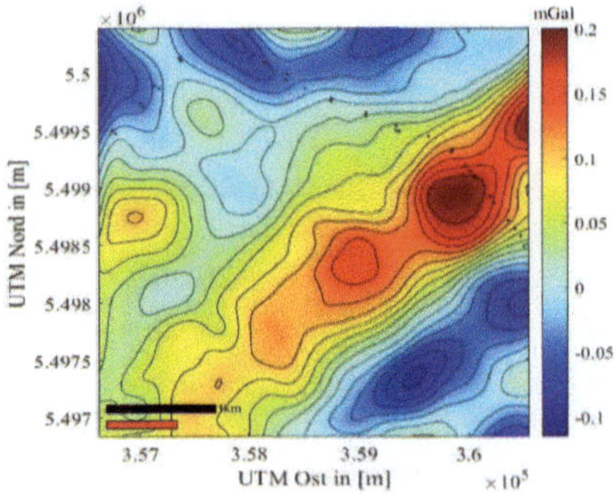

Gravitational anomaly scale $j = 8$

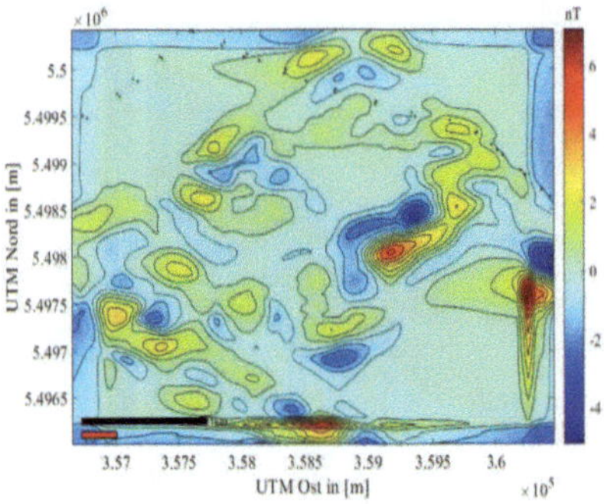

Magnetic anomaly scale $j = 9$

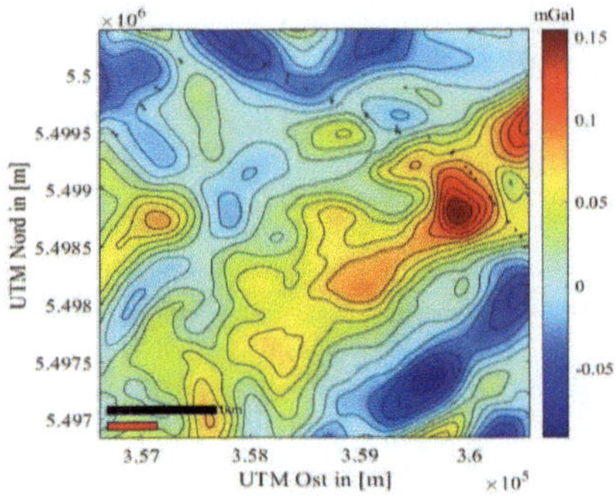

Gravitational anomaly scale $j = 9$

Fig. 8.10 Comparison of the magnetic field anomaly (left) and the gravitational field anomaly (right) in the Nohfelden test area, continued.

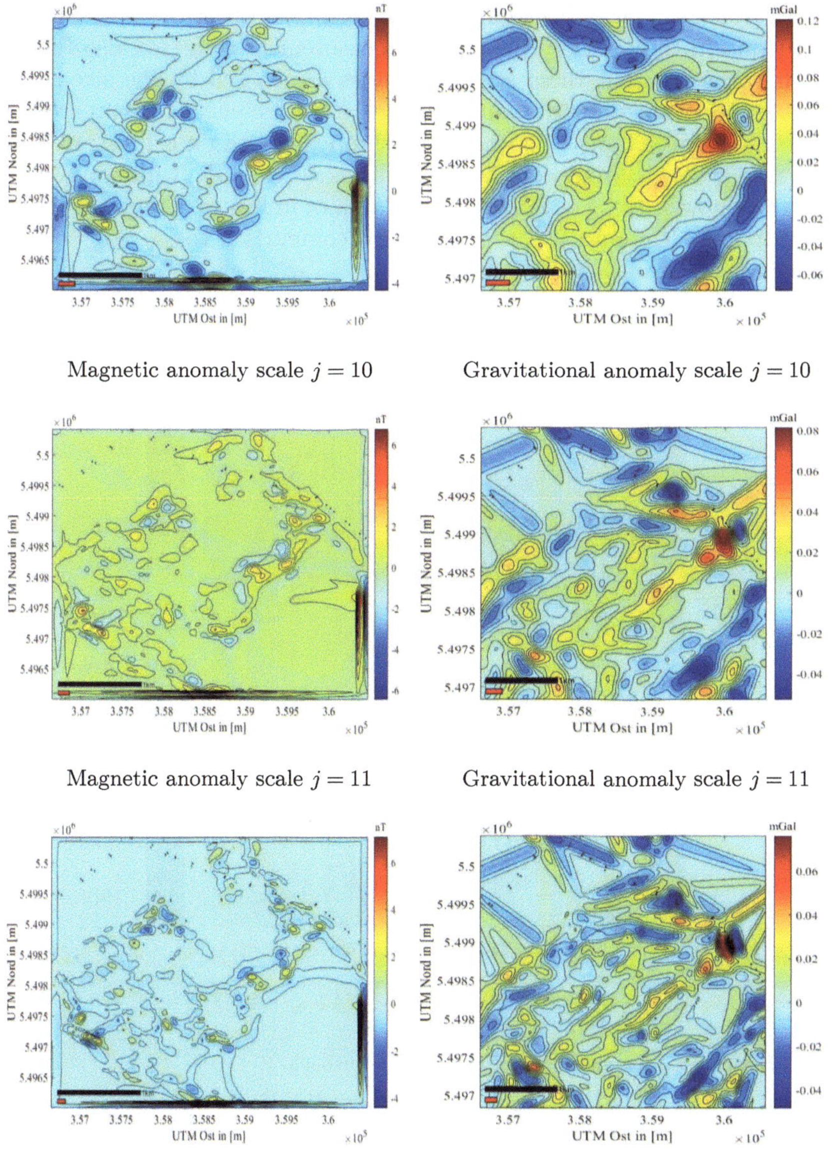

Magnetic anomaly scale $j = 10$ Gravitational anomaly scale $j = 10$

Magnetic anomaly scale $j = 11$ Gravitational anomaly scale $j = 11$

Magnetic anomaly scale $j = 12$ Gravitational anomaly scale $j = 12$

Fig. 8.11 Comparison of the magnetic field anomaly (left) and the gravitational field anomaly (right) in the Nohfelden test area, continued.

Surface interpretation of the Saarland test area Burbach. The combined evaluation of all available real data provides locally promising results. Figure 8.12 shows the magnetic field anomalies measured within SYSEXPL in the test area Burbach. There it can already be seen that the stronger anomalies are all found in the southern part of the measurement area (see also Figure 8.13). It turns out that these large anomalies are caused by the strong inhomogeneities of the industrial and built-up areas present here compared to the northern forested area. The largely anthropogenic origin is illustrated by Figure 8.14, where a strong correlation to the existing roads (white lines) and the existing power line (green line), whose area of influence is represented by the transparent white area, can be seen, for comparison the geology of the Burbach area is also shown (Figure 8.15) .

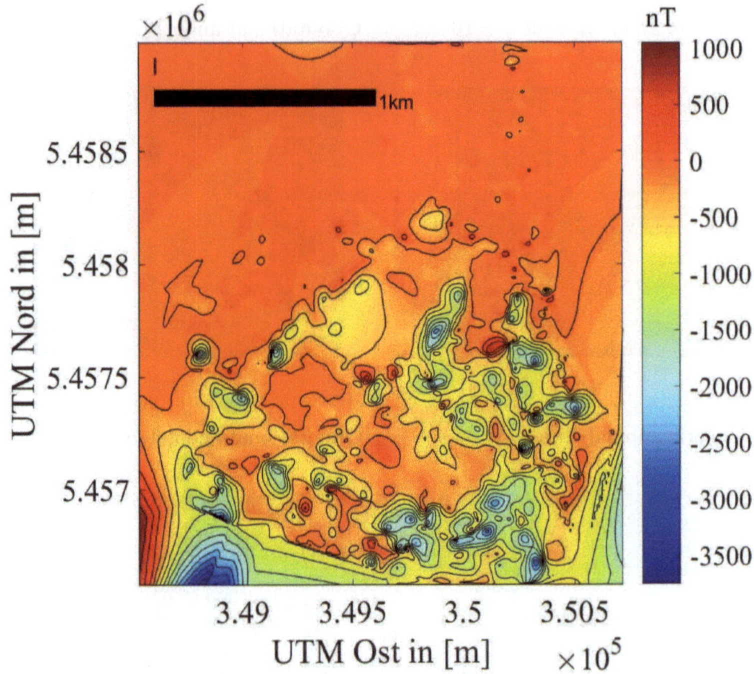

Fig. 8.12 Measured magnetic field anomalies in Burbach.

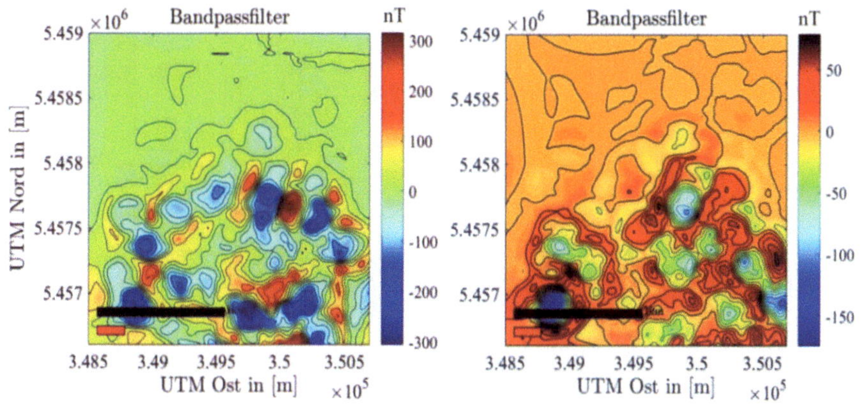

Fig. 8.13 Comparison of the decorrelations of the total intensity (left) and vertical gradient (right) anomalies at bandpass filter $j = 8$.

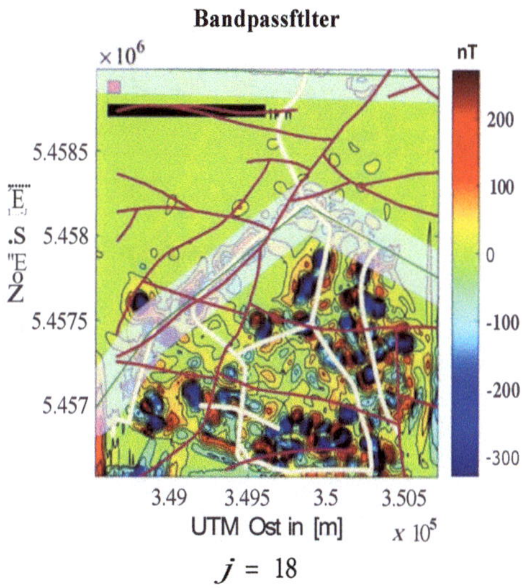

Fig. 8.14 Interpretation of the magnetic field anomaly in Burbach area.

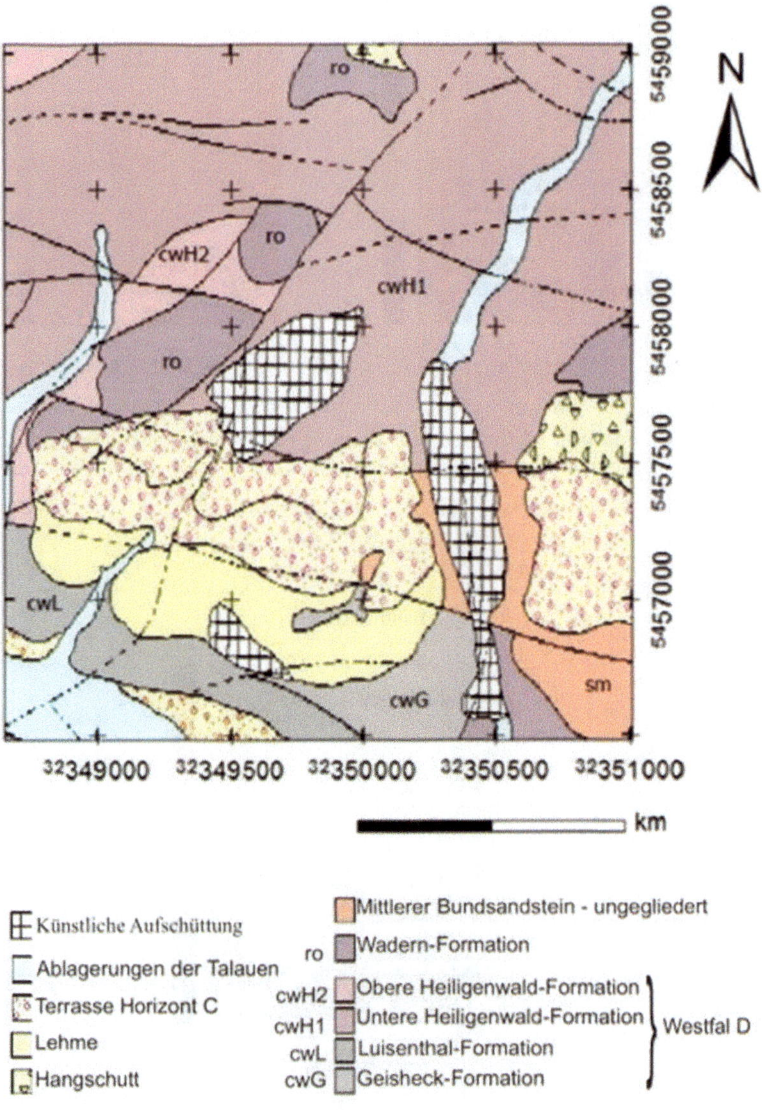

Fig. 8.15 Geology in Burbach area (see also Figure 7.5).

The decorrelation of the data in Figures 8.16 to 8.19 provides further insight into the present magnetic field anomalies.

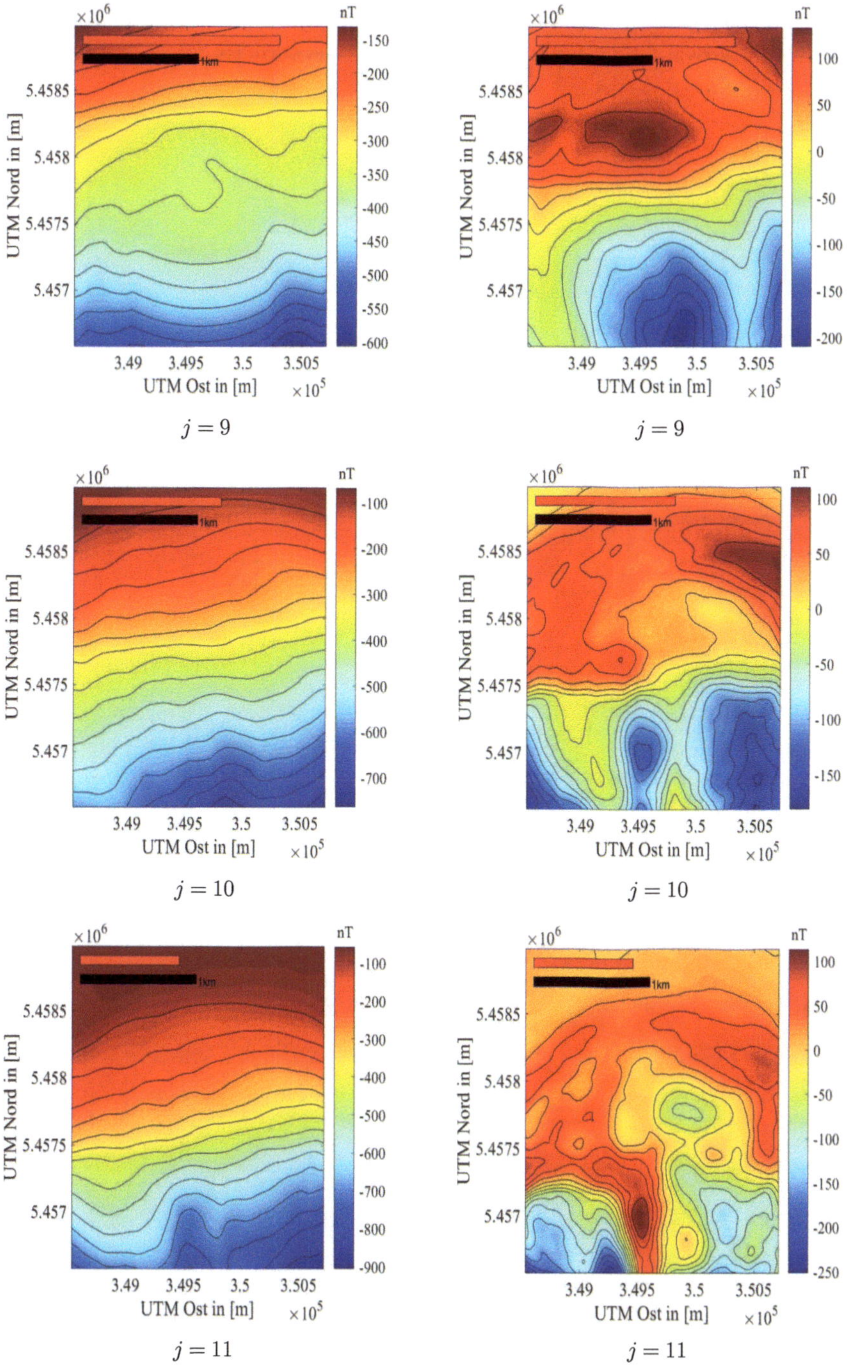

Fig. 8.16 Decorrelation of the magnetic field anomaly dataset in the Burbach area from Figure 8.12.

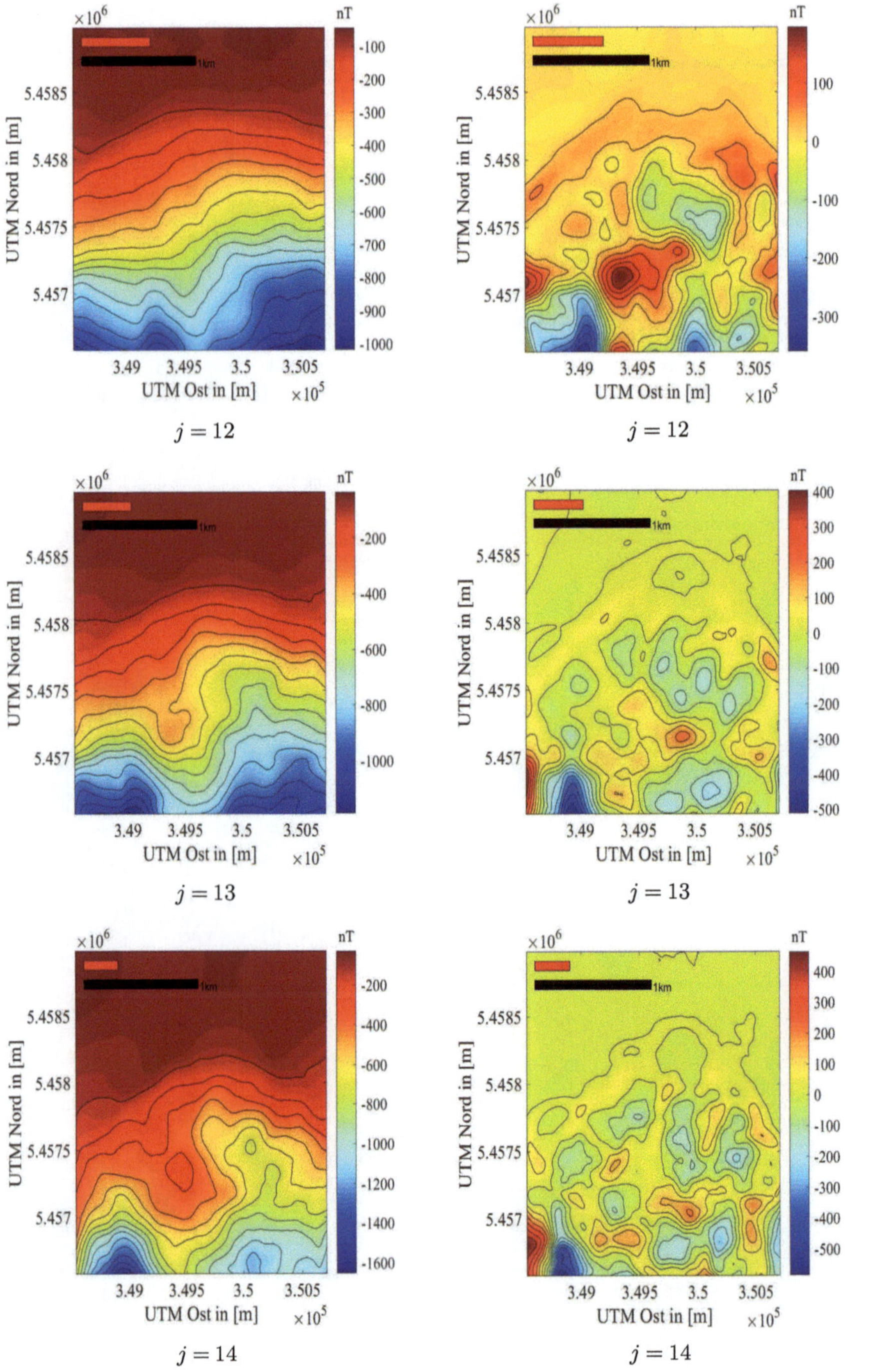

Fig. 8.17 Decorrelation of the magnetic field anomaly dataset in the Burbach area from Figure 8.12, continued.

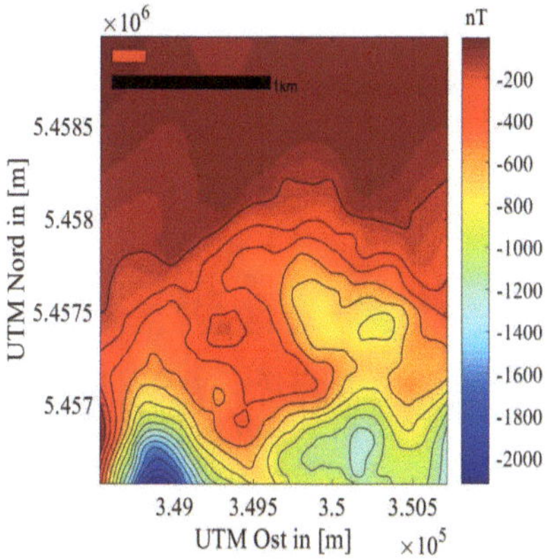

$j = 15$

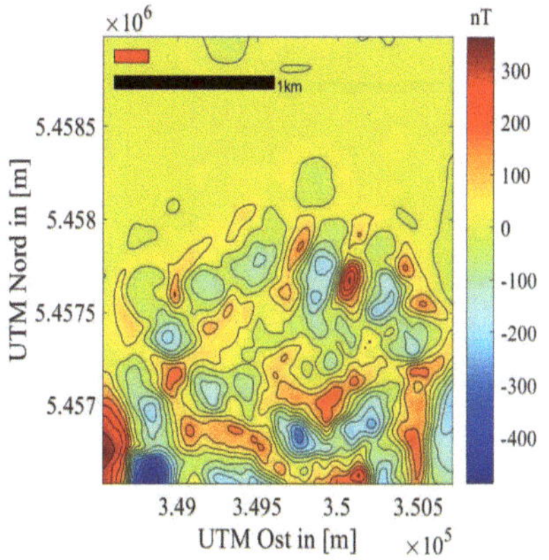

$j = 15$

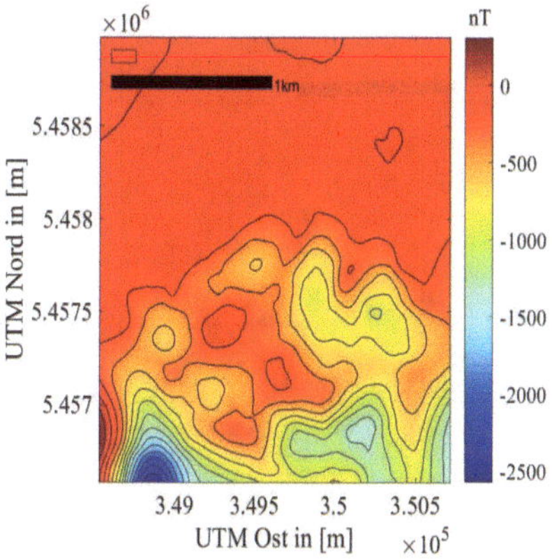

$j = 16$

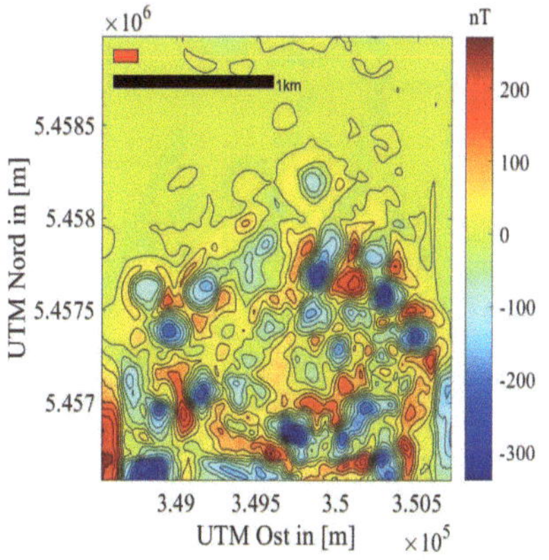

$j = 16$

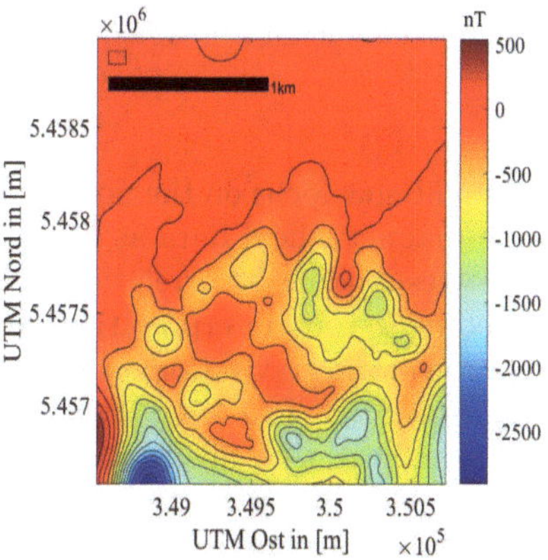

$j = 17$

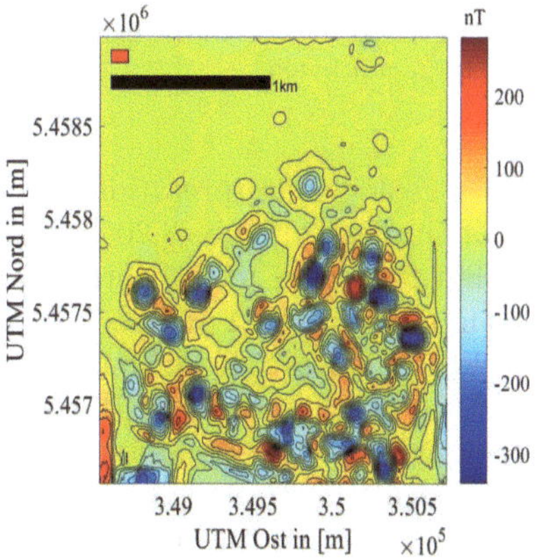

$j = 17$

Fig. 8.18 Decorrelation of the magnetic field anomaly dataset in the Burbach area from Figure 8.12, continued.

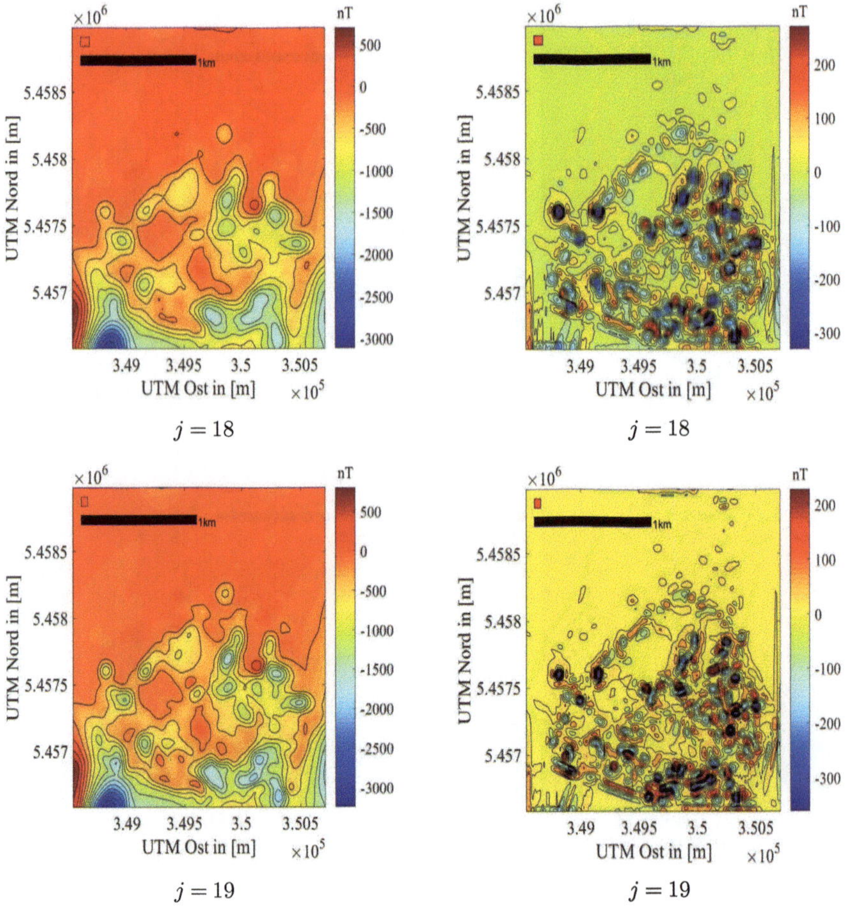

Fig. 8.19 Decorrelation of the magnetic field anomaly dataset in the Burbach area from Figure 8.12, continued.

For a geologic interpretation of the data we will now focus on the northern area of Burbach. Figure 8.20 shows the same magnetic field data as Figure 8.12, but restricted to the northern area. A corresponding decorrelation can be seen in Figures 8.21 and 8.22. Before proceeding to their interpretation, we describe in Figures 8.25 and 8.26 the decorrelation of the gravitational anomalies of the northern area of Burbach known from SPE (PtJ FKZ 0324061) (see Figure 8.23), to be able to draw a later comparison to magnetics. The interpretation of the decorrelation of magnetic field anomaly and gravity anomaly is discussed using the summary of the relevant bandpass filtering in Figure 8.27. This shows the bandpass filtering of the magnetic field anomaly to scales $j = 11$ and $j = 12$ at the top and the bandpass filtering of the gravitational anomaly to scale $j = 13$ at the bottom left. In all three images, the fault zones known from geology (cf. Figure 8.27 bottom right) are shown in red. It turns out that a geologic interpretation of the magnetic field anomaly in this area can be carried out, since the measurement area is only slightly anthropogenically influenced. The bandpass filtered magnetic field anomaly to scale $j = 11$ already shows a correlation to the existing fault zones. The area shown in red is clearly divided into two areas by a fault zone. This correlation is more clearly visible in the scale $j = 12$. The red areas can be clearly bounded by the plotted fault zones. While the magnetic field anomaly provides information about existing fault zones, the decorrelation of the gravity anomaly gives information about the existing geologic layers. The blue areas from Figure 8.27 at the lower left clearly correlate to the Wadern formations from the geology.

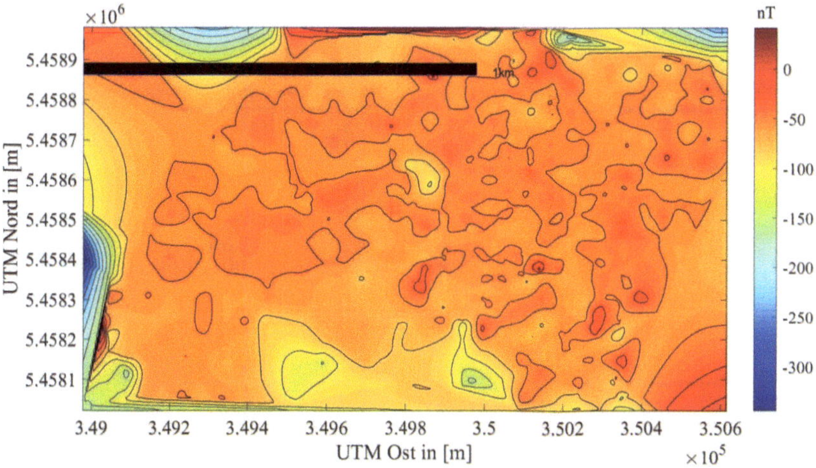

Fig. 8.20 Measured magnetic field anomalies in Burbach North.

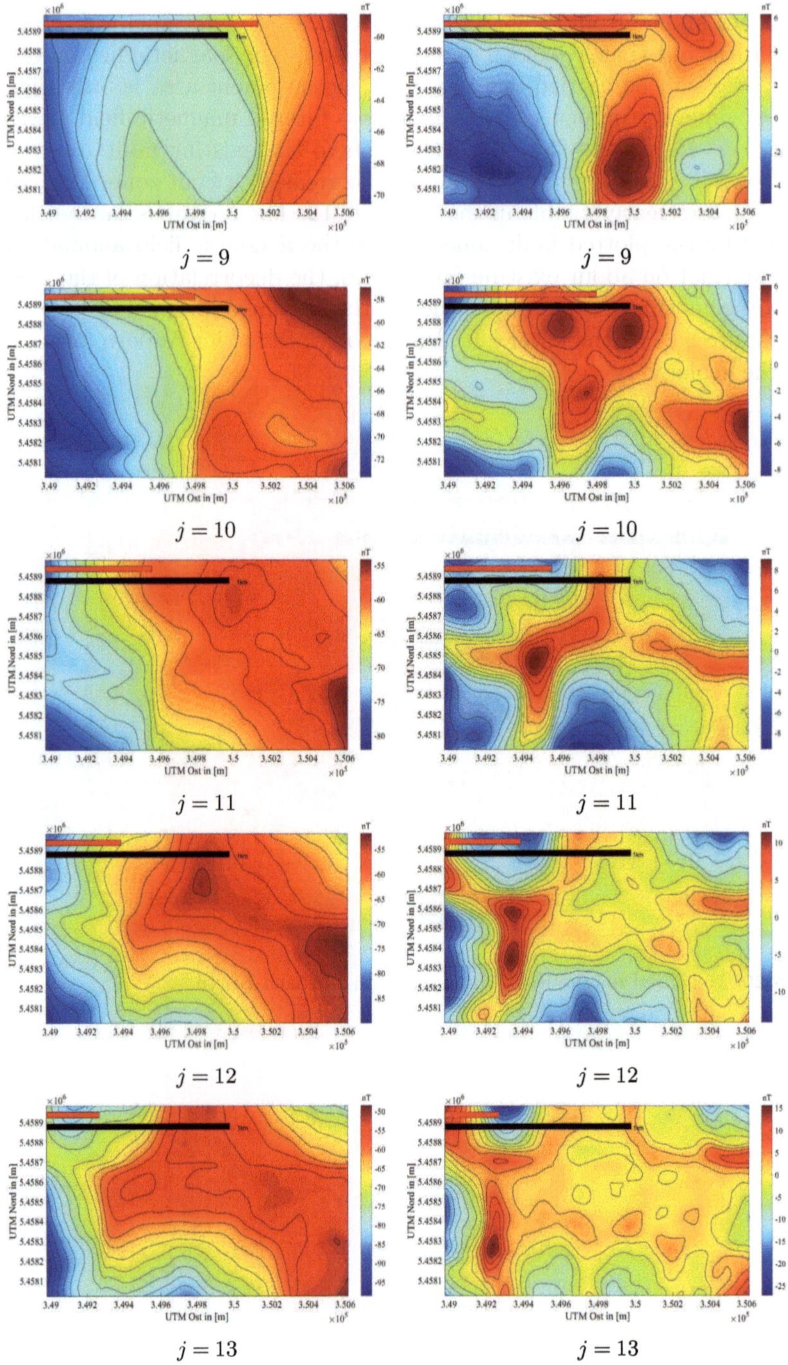

Fig. 8.21 Decorrelation of the magnetic anomaly dataset in Burbach North from Figure 8.20.

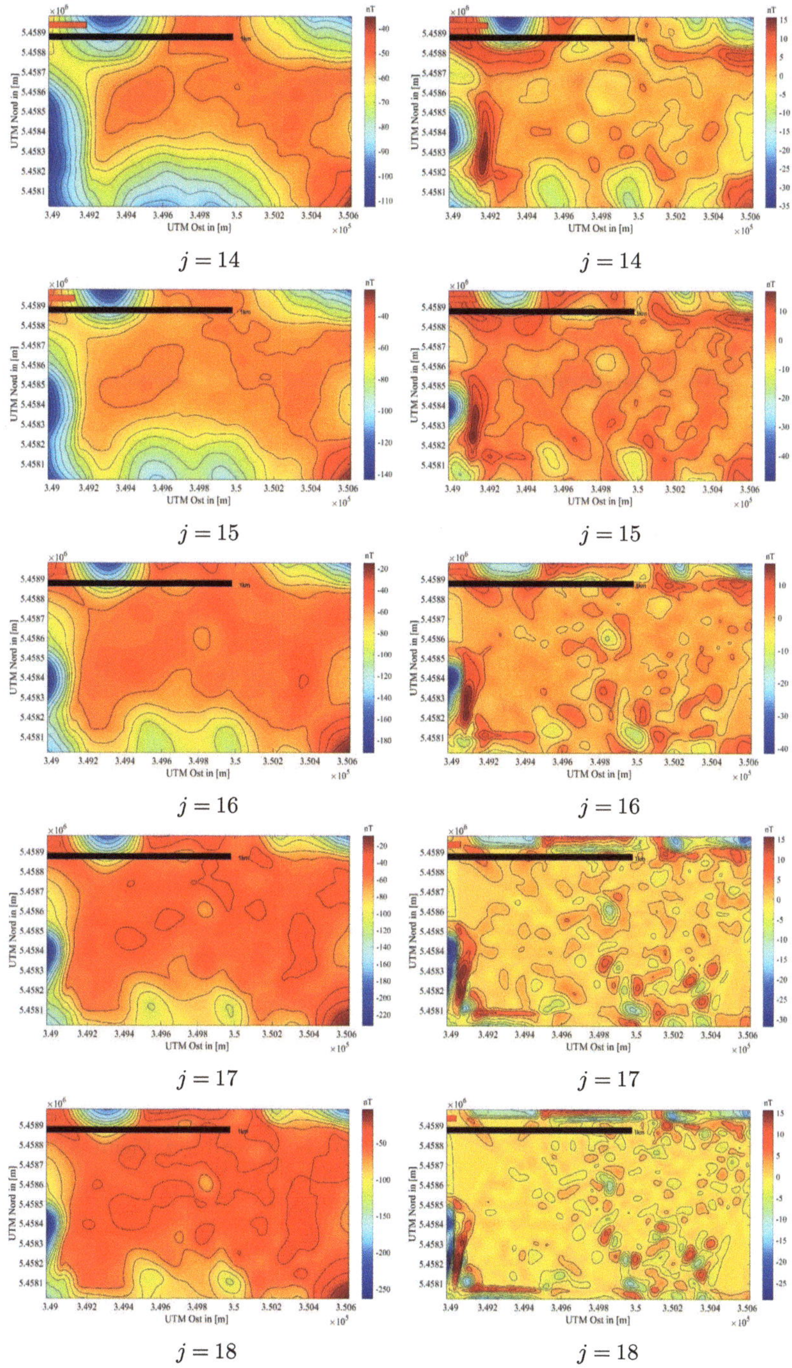

Fig. 8.22 Decorrelation of the magnetic anomaly dataset in Burbach North from Figure 8.20, continued.

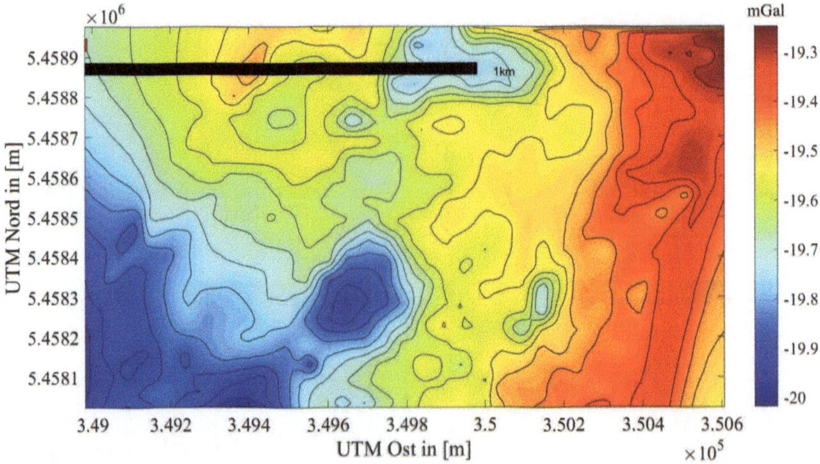

Fig. 8.23 Measured gravity field anomalies in Burbach North.

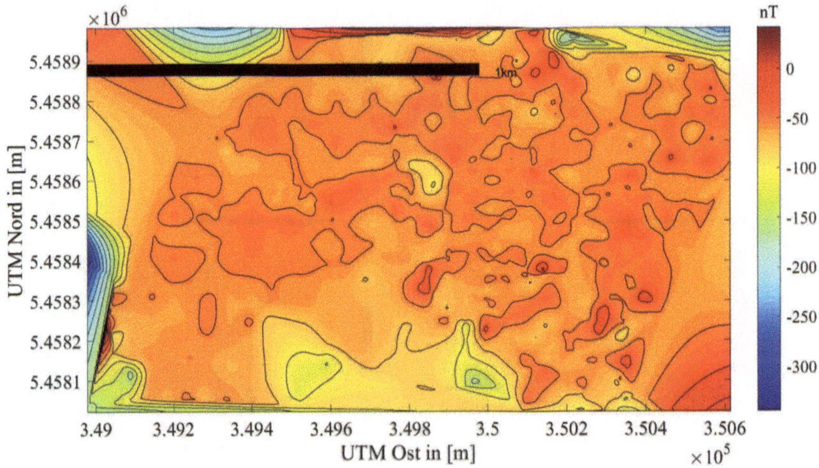

Fig. 8.24 Measured magnetic field anomalies in Burbach North (cf. Figure 8.20.

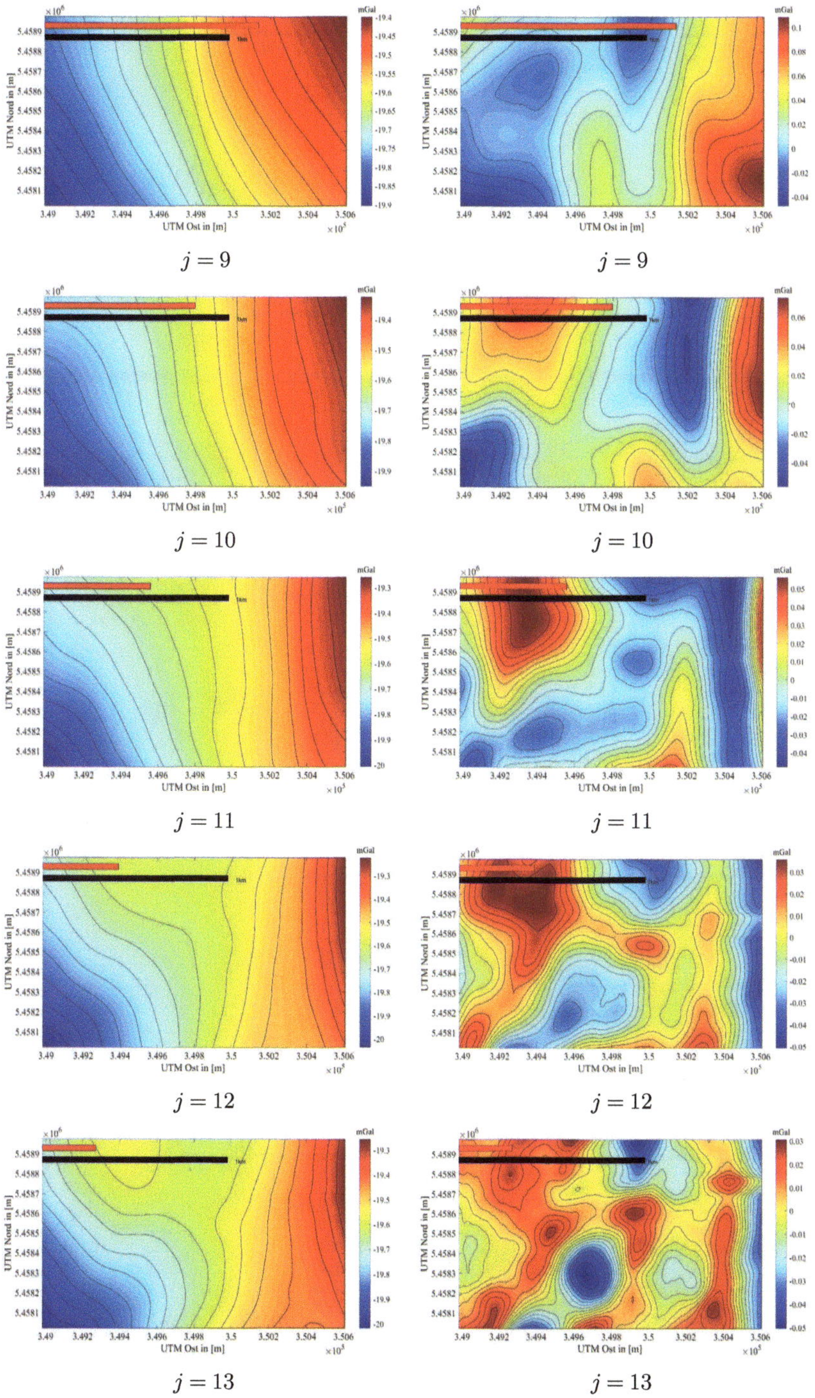

Fig. 8.25 Decorrelation of the gravity field anomaly dataset in Burbach North from Figure 8.23.

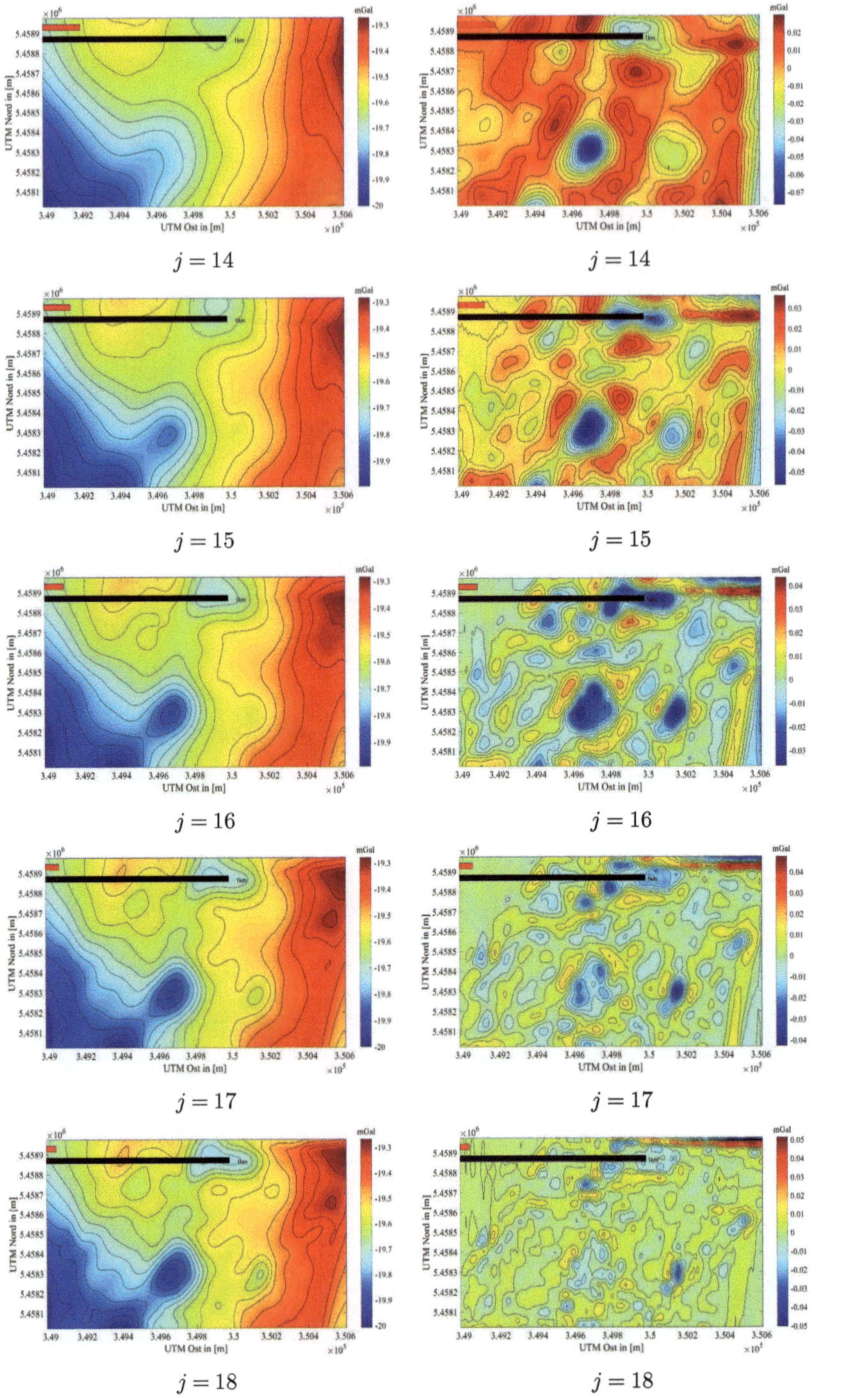

Fig. 8.26 Decorrelation of the gravity field anomaly dataset in Burbach North from Figure 8.23, continued.

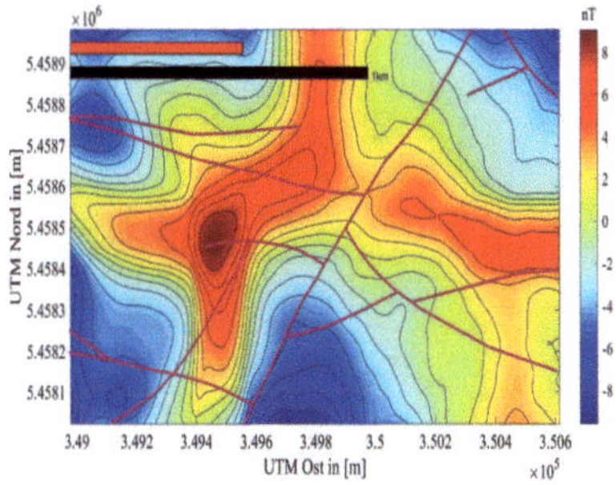

Magnetic field anomaly for $j = 11$

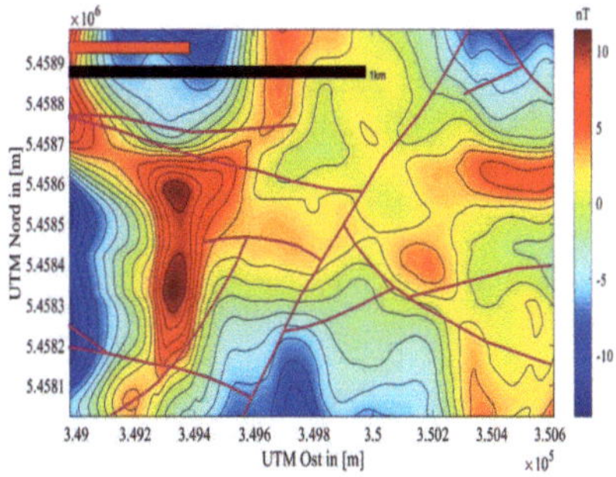

Magnetic field anomaly for $j = 12$

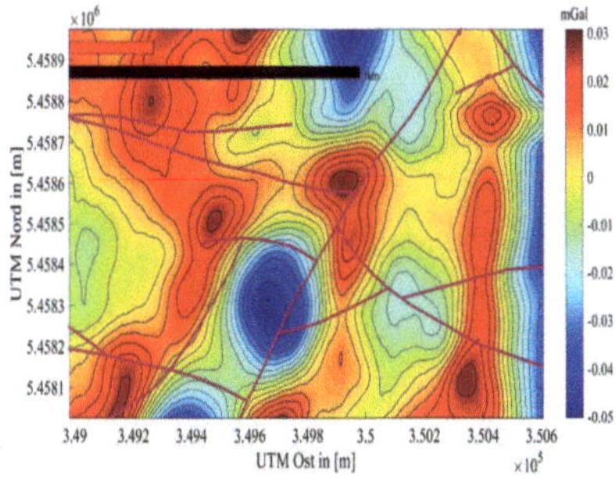

Gravitational anomaly for $j = 13$

Fig. 8.27 Comparison of the magnetic field anomaly decorrelation (top) and the gravity field anomaly decorrelation at Burbach North.

Final assessment of the results for surface analysis. A juxtaposition of the decorrelations of gravitational anomalies and anomalies of the geomagnetic total field promises successful comparison and/or supplementation possibilities with respect to the geologic interpretation on a large scale but also on a small scale. Geologic units and tectonic structures can be recognized in both decorrelations, but to different degrees. A pronounced similarity of the gravity anomalies with the magnetic field anomalies is reflected, for example, at small scales of the decorrelations in the Hunsrück-Südrand fault. This is characterized as a SW-NE-trending magnetic band structure with superimposed low-amplitude short-wavelength anomalies, and it stands out as a sharper, discontinuous SW-NE lineament in the bandpass filtering of the gravity anomalies. At higher scales, only magnetizable rocks come out in the decorrelations of the total geomagnetic field anomalies, whereas in the gravimetric decorrelations the large perturbations become more prominent. For small areas, the results are very much influenced by the data accuracy and density, as well as precise determination of the location of the measurement points.

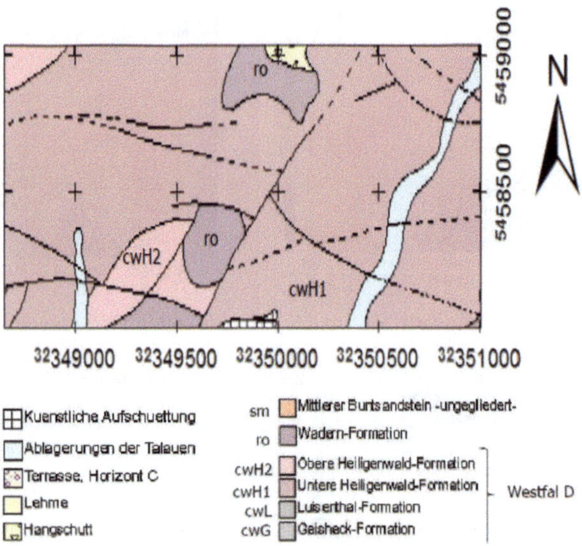

Fig. 8.28 Geology at Burbach North.

With the help of systematic exploration using gravity and magnetic field data, more information about a geologic area can be obtained than each dataset can provide on its own. In addition, systematic exploration allows mutual validation of the respective datasets.

8.2 Geologic mollifier decorrelation using depth methodologies

The goal of geomathematical depth modeling is the decorrelative representation of 3D models of the density or magnetization within the Earth's interior of the test area. In doing so, it is helpful to recall the following two aspects of using gravimetric or magnetometric information to model the subsurface:

- Due to the integral equations underlying the physical law, the information obtained by appropriate measurements is smooth. As a consequence, from gravimetric or magnetometric information one does not obtain sharp contours of the potential in the region of interest. Wavelet decorrelation, however, opens up the possibility already at the potential level to decompose the signal suitably into different signature bands and in this way to "roughent" the smoothness by band specification and to document transitions.

- A suitable redistribution of density or magnetization leaves the resulting potential unchanged. This uniqueness problem forces a certain amount of prior information from the study area for practical reference. It should be noted again that a complete reconstruction of density or magnetization from terrestrial measurements only is generally impossible because of the unavoidable "ghosts" that occur. Seismic information, downhole measurements, or similar additional information about the study area - such as susceptibility - should always be comprehensively incorporated into the modeling for this reason.

The goal of gravimetric/magnetometric depth exploration is thus the creation of a three-dimensional geologic subsurface model that is as realistic as possible as a basis for decision-making, e. g., for (geothermal) exploration, from gravitational field or magnetic field information. The macroscopic physical principles underlying this task are Newton's law of gravity (gravimetry) and Biot-Savart's law (magnetometry), which describe the determination of the potential or its functionals from a known density distribution or magnetization in a test area. In the case of deep exploration, one strives to solve the inverse problem, that is, to determine the internal density distribution or magnetization from available information on the potential. This so-called inverse problem, in which the unknown cause (density distribution, magnetization), which is not or only with difficulty accessible to direct measurement, is to be determined from the known effect (potential), has already been explained in detail.

We are interested in a discrete inversion method that is able to generate a subsurface model from nonlinear gravitational or magnetic field data, incorporating already known information about the exploration area, e. g., obtained from geologic investigations of a different nature. This additional information is intended to reduce or overcome the imponderables due to the mathematical situation of the ill-posedness of the inverse problem of gravimetry and/or magnetometry. In other words, the information already known in the test area is used to eliminate problems of uniqueness and stability in the process of inversion, in order to get as close as possible to the actual geologic structure with the calculated result. The methodology used essentially leads to a discrete nonlinear minimization problem (see also [54]). The mollifier wavelet functions with their spherical supports become scale-dependent ingredients within nonlinear systems of equations. The goal is to obtain from the superficially available measured data (geopotential intensities, i. e., amounts of gravity and/or magnetic field) in data points x_i, $i = 1, \ldots, m$, $x_i \in \mathcal{B}$, to generate a decorrelation model of the density distribution or material magnetization in nodes y_k, $k = 1, \ldots, n$, $y_k \in \mathcal{B}$, in the interior.

Nonlinear gravimetric inversion. The nonlinear inversion methodology starts with a smoothed solution approach to the scale $J_0 < J$ and computes the inversion by solving a discrete nonlinear minimization problem. For clarity, we use the abbreviation

$$(W\nabla V)^n_{\tau_{J_0},\tau_J}(x) = \int_{\mathcal{B}} \nabla_x (WG)^n_{\tau_{J_0},\tau_J}(x,y)\ \rho(y)\ dV(y), \quad x \in \mathbb{R}^3,$$

with

$$(WG)^n_{\tau_{J_0},\tau_J}(x,y) = G^n_{\tau_J}(x,y) - G^n_{\tau_{J_0}}(x,y).$$

We consider the functional

$$\sigma^{cont}_{\beta,\delta}(\rho) = \sum_{i=1}^{N} \left(\frac{|(W\nabla V)^n_{\tau_{J_0},\tau_J}(x_i)| - |\nabla \tilde{V}(x_i)|}{\beta_i} \right)^2 + \delta \int_{\mathcal{B}} (\rho(y) - \tilde{\rho}(y))^2\ dV(y),$$

where the observed intensities of the vector of gravitational perturbations ∇V

$$|\nabla \tilde{V}(x_i)| = |\nabla \tilde{V}(x)||_{x=x_i}, \quad x_i \in \mathbb{R}^3,\ i = 1,\ldots,N,$$

are given and $\tilde{\rho}$ is an observed prior information about the mass density distribution ρ in $\mathcal{B}$.

To arrive at numerical results, we need to discretize $\sigma^{cont}_{\beta,\delta}(\rho)$ by applying a suitable cubature formula. This leads to $\sigma^{discr}_{\beta,\delta}(\rho_1,\ldots,\rho_N)$ given by

$$\begin{aligned} &\sigma^{discr}_{\beta,\delta}(\rho_1,\ldots,\rho_N) \qquad (8.1)\\ &= \sum_{i=1}^{N} \frac{1}{\beta_i^2} \left(\left| \sum_{k=1}^{M} w_k \nabla_x \left((WG)^n_{\tau_{J_0},\tau_J}(x,y_k)|_{x=x_i} \right) \rho_k \right| - |\nabla \tilde{V}(x_i)| \right)^2 \\ &+ \delta \sum_{k=1}^{M} w_k \, |\rho_k - \tilde{\rho}(y_k)|^2 . \end{aligned}$$

Here w_k, y_k, $k = 1,\ldots,M$, are the known weights and the given nodes of the cubature formula, respectively, and

$$\rho_k = \rho(y_k), \quad k = 1,\ldots,M,$$

the unknown values of the mass density for the points y_k, $k = 1,\ldots,M$, from $\mathcal{B}$.

$(\tilde{\rho}(y_1),\ldots,\tilde{\rho}(y_M))^T$ thus stands for the known pre-information of the unknown density vector $(\rho(y_1),\ldots,\rho(y_M))^T$ and $|\nabla \tilde{V}(x_i)|$, $i = 1,\ldots,N$, characterizes the known intensities at points $x_i \in \mathbb{R}^3$.

The values β_i, $i = 1,\ldots,N$, are fitted to the standard deviations of the observed data values $|\nabla\ \tilde{V}(x_i)|$, $i = 1,\ldots,N$. The positive parameter δ serves as a smoothing control, i. e., the larger δ is chosen, the stronger the influence of the a priori information $(\tilde{\rho}(y_1),\ldots,\tilde{\rho}(y_M))^T$ on the unknown values to be calculated $(\rho(y_1),\ldots,\rho(y_M))^T$.

The expression $(\sigma_{\beta,\delta}^{discr}(\rho_1,\ldots,\rho_M))^2$ can be interpreted as the square of the $(N+M)$-dimensional vector

$$\begin{pmatrix} \frac{1}{\beta_1}\left|\sum\limits_{k=1}^{M} w_k \nabla\left(G^n_{\tau_J}(x_1,y_k) - G^n_{\tau_{J_0}}(x_1,y_k)\right)\rho_k\right| - |\nabla\tilde{V}(x_1)| \\ \vdots \\ \frac{1}{\beta_N}\left|\sum\limits_{k=1}^{M} w_k \nabla\left(G^n_{\tau_J}(x_N,y_k) - G^n_{\tau_{J_0}}(x_N,y_k)\right)\rho_k\right| - |\nabla\tilde{V}(x_N)| \\ \sqrt{\delta}\ \sqrt{w_1}\,|\rho_1 - \tilde{\rho}(y_1)| \\ \vdots \\ \sqrt{\delta}\ \sqrt{w_M}\,|\rho_M - \tilde{\rho}(y_M)| \end{pmatrix}.$$

The unknown coefficients are defined to minimize the expression, i. e.,

$$(\sigma_{\beta,\delta}^{discr}(\rho_1,\ldots,\rho_M))^2 \to \min$$

holds in the set of all M-tuples $(\rho_1,\ldots,\rho_M)^T$. It should be emphasized that the result of the minimization procedure depends essentially on the "smoothing parameter" δ.

The formulas for the nonlinear gravimetric inversion result from numerical approaches of the dissertation [28]. In the form given above they can be found in [54].

Test area Burbach. Depth modeling using gravimetric inversion was applied in the Burbach test area. An initial density model for the iteration process was obtained from all available information about the geology of the test area. This includes information from the Saarland mining industry as well as from boreholes drilled in the nearby area.

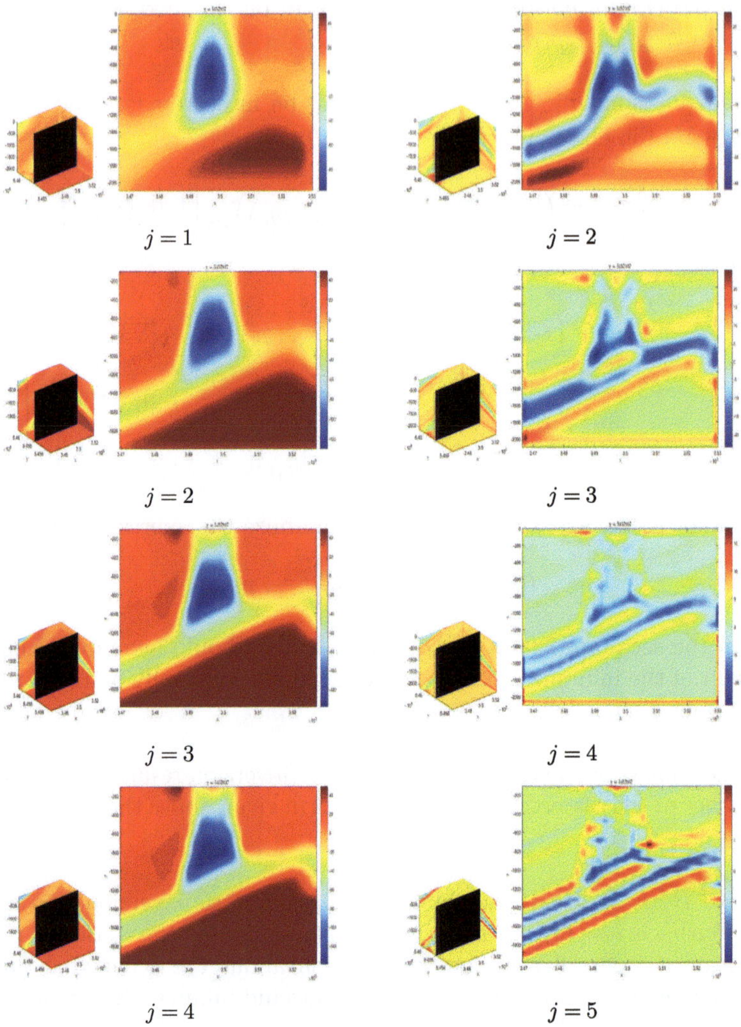

Fig. 8.29 Decorrelation of the density distribution inside Burbach obtained from measured gravity field anomalies under some use of geologic preinformation about the Saar fault (lowpass components on the left, bandpass components on the right).

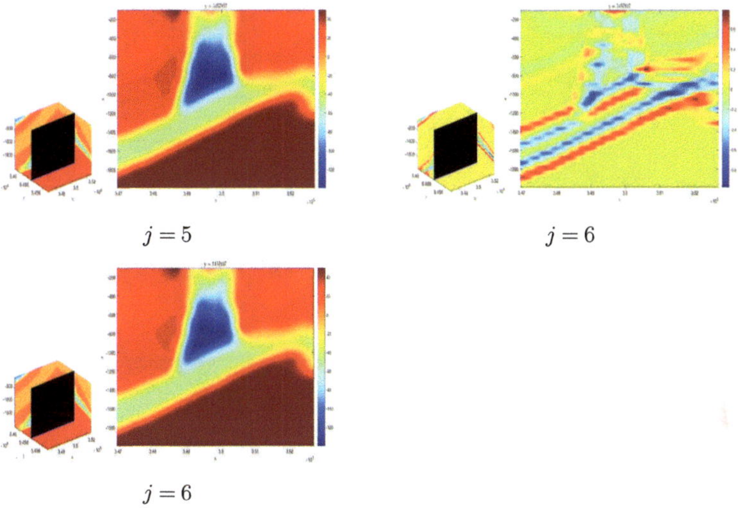

Fig. 8.30 Decorrelation of the density distribution inside Burbach obtained from measured gravity field anomalies under some use of geologic preinformation about the Saar fault (lowpass components on the left, bandpass components on the right), continued.

Figures 8.29 and 8.30 show the decorrelation of the density in the subsurface in the Burbach test area. The different geologic layers can be delineated based on their density differences, and layers with high density contrast can be read directly from the Figures 8.29 and 8.30. For example, the location of the Saar fault can be readily identified.

By means of the decorrelation, the inversion result can be broken down further for interpretation purposes, so that the exact position of the layer boundaries in the calculated model becomes apparent (see Figure 8.31). Here, the boundary between two layers is generally given where a sign change occurs. Therefore, for better interpretability, in Figure 8.31 areas close to zero are shown in white.

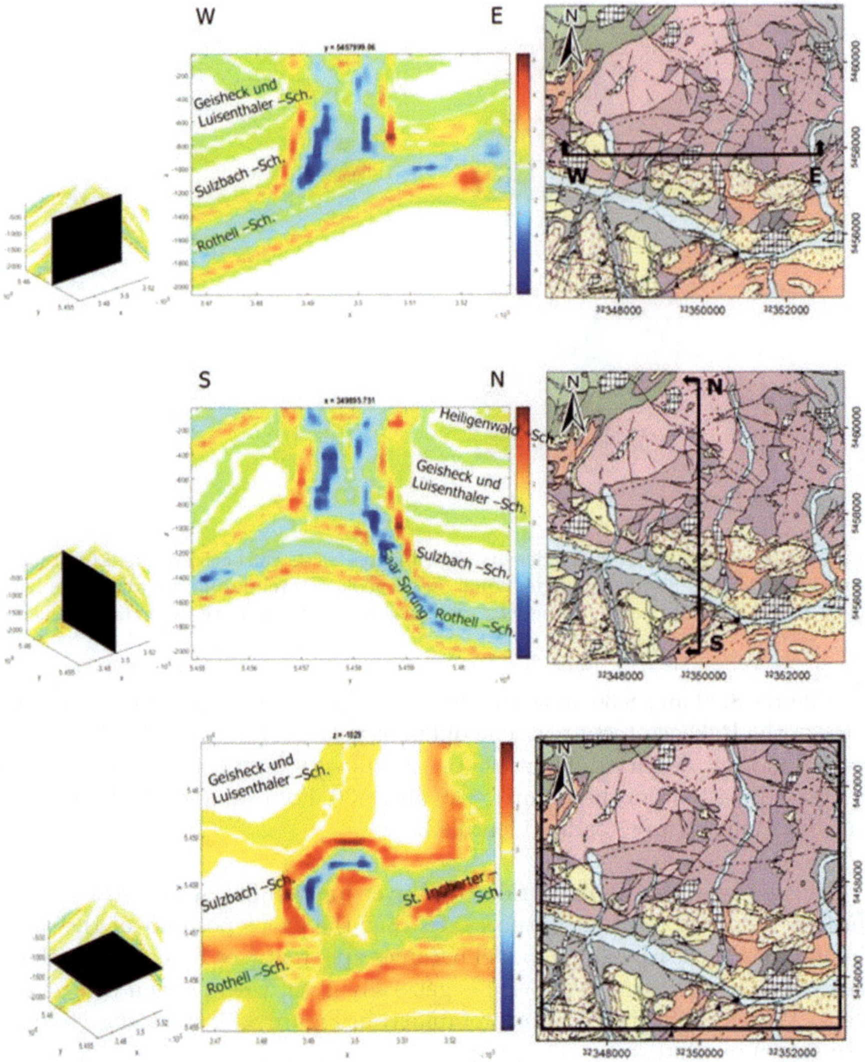

Fig. 8.31 Layer boundary analysis of the inversion result from Figures 8.29 and 8.30 and location of the section in the geologic map

In all three profiles of the Figure 8.31 the safety pillar of the Burbach repair company ("Burbach Ausbesserungswerk") and the deeper lying Rothell strata can be clearly distinguished due to the higher density contrast. Furthermore, the course of the Saar fault can be seen in profiles W-E and S-N. In the other geologic layers, the densities of the rock formations are close to each other, so that the boundary between them is not so clear; for example, the boundary Geisheck/Sulzbacher layers is between the yellowish green and light green. In the bottom image of Figure 8.31, which represents a depth of about 1 000 m, the circular outline of the safety pier of the Burbach repair company is visible because the coal seams have not been mined in this area and thus the same layer has a different density. The St. Ingbert strata, which seem to occur in the Rothell strata, are prominent here. In this area southwest of the Saar fault, the strata are folded. Obviously, the geologic layers are better delineated in depth, the greater the difference in density between them.

Nonlinear magnetometric inversion. In analogy to gravimetry, for nonlinear inversion we deal with a mollifier methodology which considers the susceptibility χ instead of the magnetization distribution m. The corresponding formulas for inversion for magnetometry can be found in [34]. We start with an abbreviation

$$(Wb)^n_{\tau_{J_0},\tau_J}(x) = \int_{\mathcal{B}} \nabla_x (W\gamma)^n_{\tau_{J_0},\tau_J}(x,y) \cdot b(y)\ \chi(y)\ dV(y), \quad x \in \mathbb{R}^3, \tag{8.2}$$

where

$$(W\gamma)^n_{\tau_{J_0},\tau_J}(x,y) = \gamma^n_{\tau_J}(x,y) - \gamma^n_{\tau_{J_0}}(x,y). \tag{8.3}$$

We consider the “smoothing functional”

$$\sigma^{\text{cont}}_{\beta,\delta}(\chi) = \sum_{i=1}^{N} \left(\frac{|(Wb^n_{\tau_{J_0},\tau_J}(x_i)| - |\tilde{b}(x_i)|}{\beta_i} \right)^2 + \delta \int_{\mathcal{B}} (\chi(y) - \tilde{\chi}(y))^2\ dV(y), \tag{8.4}$$

where the observed intensities $|\tilde{b}(x_i)|, x_i \in \mathbb{R}^3$, $i = 1,\ldots,N$, of the field b are given and $\tilde{\chi}$ is an observed prior information about the susceptibility χ in $\mathcal{B}$.

We discretize $\sigma^{\text{cont}}_{\beta,\delta}(\chi)$ by applying a suitable cubature formula leading to the functional $\sigma^{\text{discr}}_{\beta,\delta}(\chi_1,\ldots,\chi_N)$ given by

$$\begin{aligned} &\sigma^{\text{discr}}_{\beta,\delta}(\chi_1,\ldots,\chi_N) \\ &= \sum_{i=1}^{N} \frac{1}{\beta_i^2} \left(\left| \sum_{k=1}^{L} w_k \left(\nabla_x (W\gamma)^n_{\tau_{J_0},\tau_J}(x,y_k)|_{x=x_i} \cdot b(y_k) \right) \chi_k \right| - \left| \tilde{b}(x_i) \right| \right)^2 \\ &\qquad + \delta \sum_{k=1}^{L} w_k\, |\chi_k - \tilde{\chi}(y_k)|^2 . \end{aligned} \tag{8.5}$$

Here w_k, y_k, $k = 1,\ldots,L$, are the known positive weights and the prescribed nodes of the cubature formula, respectively, and $\chi_k = \chi(y_k)$, $k = 1,\ldots,L$, are the unknown susceptibility values for the points y_k, $k = 1,\ldots,L$, in $\mathcal{B}$.

In other words, $(\tilde{\chi}(y_1),\ldots,\tilde{\chi}(y_L))^T$ represents the known pre-information of the unknown susceptibility vector $(\chi(y_1),\ldots,\chi(y_L))^T$, and $|\tilde{b}(x_i)|$, $i = 1,\ldots,N$, characterizes the known (observed) intensities of the field vectors $|b(x_i)|$, $i = 1,\ldots,N$, in the points $x_i \in \mathbb{R}^3$.

The values β_i, $i = 1,\ldots,N$, are fitted to the standard deviations of the observed data values $|\tilde{b}(x_i)|$, $i = 1,\ldots,N$. The positive parameter δ serves as a smoothing control, i. e., the larger δ is chosen, the stronger the influence of the a priori information $(\tilde{\chi}(y_1),\ldots,\tilde{\chi}(y_L))^T$ on the unknown values $(\chi(y_1),\ldots,\chi(y_L))^T$ to be calculated.

Indeed, it is not difficult to verify that the expression $(\sigma^{\text{discr}}_{\beta,\delta}(\chi_1,\ldots,\chi_N))^2$ can be identified as the square of the $(N+L)$-dimensional vector

$$\begin{pmatrix} \frac{1}{\beta_1}\left|\sum\limits_{k=1}^{L} w_k \left(\nabla_x (W\gamma)^n_{\tau_{J_0},\tau_J}(x,y_k)|_{x=x_1}\cdot b(y_k)\right)\chi_k\right| - |\tilde{b}(x_1)| \\ \vdots \\ \frac{1}{\beta_N}\left|\sum\limits_{k=1}^{L} w_k \left(\nabla_x (W\gamma)^n_{\tau_{J_0},\tau_J}(x,y_k)|_{x=x_N}\cdot b(y_k)\right)\chi_k\right| - |\tilde{b}(x_N)| \\ \sqrt{\delta}\,\sqrt{w_1}\,|\chi_1 - \tilde{\chi}(y_1)| \\ \vdots \\ \sqrt{\delta}\,\sqrt{w_L}\,|\chi_L - \tilde{\chi}(y_L)| \end{pmatrix}. \tag{8.6}$$

The unknown coefficients are determined such that $(\sigma^{\text{discr}}_{\beta,\delta}(\chi_1,\ldots,\chi_L)^2$ is minimized in the set of all L-tuples $(\chi_1,\ldots,\chi_L)^T$. It should be emphasized that the result of the minimization procedure depends on the "smoothing parameter" δ.

As is well-known, the smoothing wavelet

$$x \mapsto (W\gamma)^n_{\tau_{J_0},\tau_J}(x,y) = \gamma^n_{\tau_J}(x,y) - \gamma^n_{\tau_{J_0}}(x,y), \quad x \in \mathcal{B}, \tag{8.7}$$

has a sphere as support whose width is determined by the scale index J_0 and thus, as already explained, has a decisive influence on the desired depth of exploration. In accordance with our model specification, the scale index $J > J_0$ should be chosen as large as possible.

The nonlinear minimization problem $(\sigma^{\text{discr}}_{\beta,\delta}(\chi_1,\ldots,\chi_L))^2 \to \min$ can be computed with standard numerical analysis solvers as for the gravimetric case.

Test area Burbach. As in the case of gravimetry, depth modeling of magnetometry was carried out in the Burbach test area. Information about the magnetization in the subsurface was available in the form of measured values about the susceptibility of the layers. A 3D susceptibility model for the Burbach area was developed based on core measurements from the Fürstenhausen-Ost well. The underlying layer boundaries were adapted to the calculated layer boundaries from the gravimetric inversion.

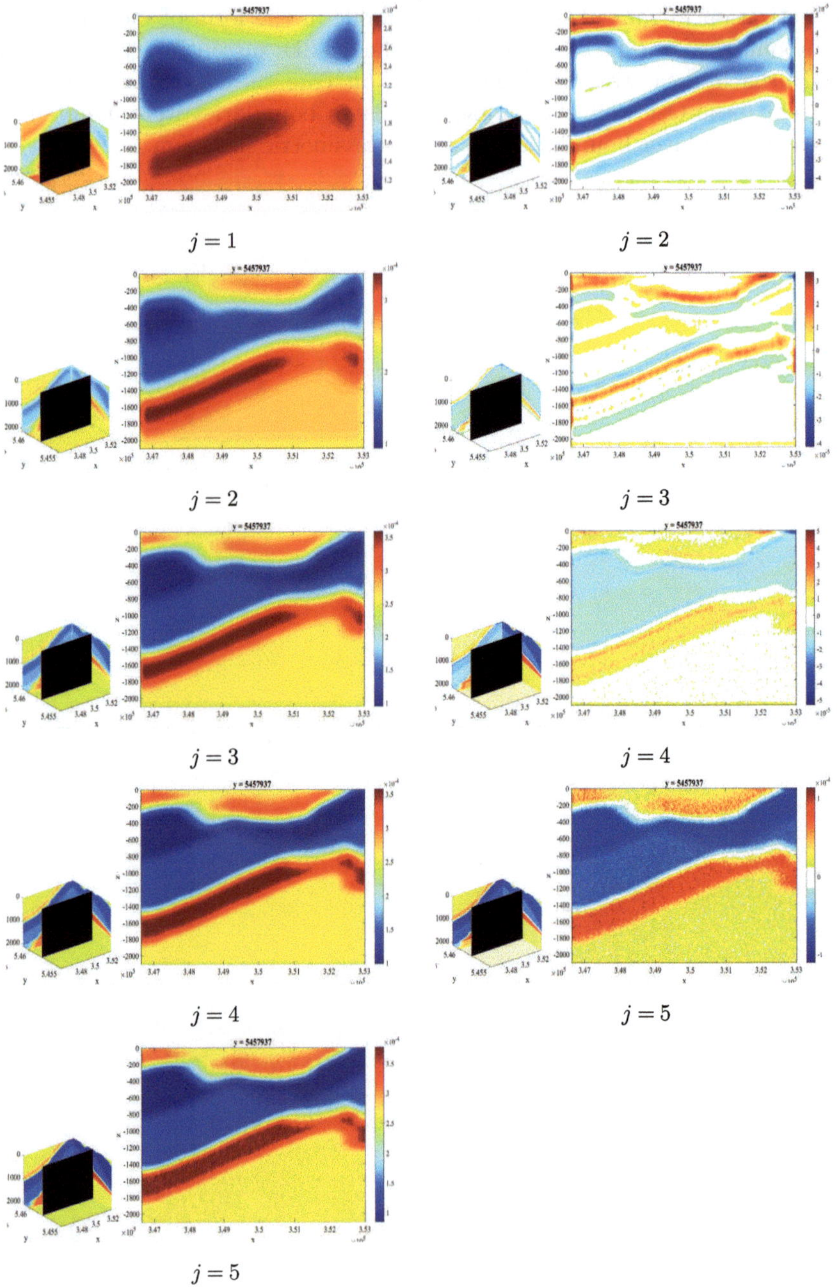

Fig. 8.32 Susceptibility decorrelation in lowpass (left) and bandpass (right) components for the test area Burbach.

It was found that the susceptibility measurements in the drill core are strongly noisy. This was taken into account accordingly in the model as good as possible.

All in all, it can be concluded that the developed decorrelation techniques can process information of a susceptibility model to facilitate interpretation. Figure 8.32 represents the decorrelation of the model. Specifically, bandpass filtering to scale $j = 3$ provides the layer transitions within the model. Band-pass filtering to scale $j \geq 4$ contains the corresponding noise components of the model, so a denoised susceptibility model can be found in the lowpass filtering to scale $j = 4$.

Chapter 9

Conclusions and outlook

Gravity and magnetic field inversion for the determination of density distribution and material magnetization in the subsurface, respectively, require a special form of regularization, that can be specifically described by inverse mollifier problems of potential theory. SPE (FKZ 0324061) and SYSEXPL (FKZ 03EE4002A) have developed new linear as well as nonlinear mollifier resolution methods for the decorrelation of geologic structures and published them in the publications [34, 54, 55, 56] to the international scientific community. A special aspect is that an inverse problem such as gravimetry, magnetometry, etc. cannot obtain uniqueness of an interior geologic structure from only terrestrial data (see Figure 3.1). Any regularization method, i. e., also the mollifier multiscale methodology presented in this work, additionally requires some prior information from the interior of the area under investigation.

Nonetheless, under reasonable data availabilty, the specific intent of this report was to show that the geomathematical treatment of gravimetric and magnetometric measurements in the desired mollifier decorrelation is able to provide the detection of the interfaces and transition levels of geologic formations due to their high-frequency signal character compared to geologically homogeneous formations, which is really a step forward in geothermal exploration.

9.1 Conclusions

The inversion of gravito/magneto-based local potential bands resulting from only local data to geologically based local contrast bands by means of wavelets is geophysically realistic in a Newtonian context by applying Poisson's equa-

W. Freeden, H. Nutz, *Exploratory Potential Methods in Geothermal Power Generation*, Lecture Notes in Geosystems Mathematics and Computing,
https://doi.org/10.1007/978-3-031-54412-5_9

tion (cf. [34, 54, 55, 62]). A suitable differential equation relates potential values pointwise to the contrast function via an associated operator derivative. This means at the same time that calculations on the potential level also yield results on the contrast distribution level.

The result of inverse gravimetry as well as magnetometry is thus to reveal geologic information in formation bands, whose mutual delimitations to each other become more and more sharply drawn with increasing scale and are, thus, responsible for the detection of formation transitions and discontinuities, which are so extremely important in exploration.

In short, within a mollifier multiscale decomposition or decorrelation process, we are able to extract detailed information that is helpful in geologic interpretation. The decorrelation method provides simultaneous results for potential and contrast distribution functions.

The novel, physically motivated mollifier multiscale modeling can be realized on the one hand as a vertical ("downward calculation") and on the other hand as a horizontal ("terrestrial surface refinement") zooming-in approximation methodology. The depth multiscale decorrelation is performed simultaneously for potential and contrast function and provides geologic formation interfaces via the decorrelation capability of wavelets.

Two alternative methods for depth-wise multiscale decorrelation for both gravimetry and magnetometry have been developed and tested, namely mollifier spline and wavelet methods, implementing the numerically advantageous principle of site localization and resulting numerical economy. The two methods are independent of each other and can contribute in this way by mutual validation to a risk reduction in each case with the sole use of gravimetry, magnetometry on the one hand and in their synopsis on the other hand in the exploration. Depths of geologic formations can also be determined by the specific use of artificially introduced faults (e. g., in point potentials of predetermined depth) and have already been tested in the Bavarian Molasse Basin (PtJ project SPE (PtJ FKZ 0324061)). Similar multiscale methods already exist for seismics (PtJ project GEOFÜND (PtJ FKZ 0325512A)).

The following advantages of decorrelative gravimetry/magnetometry can be listed for today's exploration:

- By using satellite data, a "satellite-only" model based on gradiometry (as described, for example, in [52, 70, 74]) can be incorporated as a trend.
- Gravimetric/magnetometric measurement methods are on the rise (realizing previously unknown measurement accuracies).

- Multiscale modeling can be realized as vertical (downward calculation) and horizontal (terrestrial surface refinement) zooming-in approximation methodology.
- Multiscale decorrelation with respect to the surface (terrestrial surface decorrelation) enables the improvement of geologic maps (cf., e. g., [24, 55]). It contains valuable pre-information for depth detection (cf. [34, 54]).
- Multiscale modeling with respect to depth (downward calculation) is performed numerically efficiently simultaneously with mollifier methods for potential and contrast distribution function.
- Multiscale decorrelation with respect to depth (downward calculation) provides geologic formation interfaces and indications of discontinuities.
- Two alternative methods each for depth multiscale decorrelation (risk reduction of the use of gravimetry/magnetometry) are available and applicable (cf. [34, 54]).
- Depth detection of geologic formations by artificial gravito/magneto faults (e. g., in points) is a formidable component of numerical exploration that allows accurate indications of location (cf., e. g., [32, 33]).
- Kindred multiscale methods for seismic are used for additional risk reduction by synopsis of all methods.

It may be stated: The geomathematics of the TU Kaiserslautern as well as the various contributions of the company CBM, Bexbach, have been able to successfully apply the different but mutually complementary and validating methods in the projects GEOFÜND (PtJ FKZ 0325512A), SPE (PtJ FKZ 0324061), and SYSEXPL (PtJ FKZ 03EE4002A) for reflection seismics, gravimetry and magnetics and in their combination. Thus, a basis was created to enable a systematic and scientifically based evaluation and reduction of risk for a potential investor with regard to geothermally relevant information about the subsurface. Of course, it remains to be mentioned once again that gravimetry and magnetometry can be used as exploration methods not only in anthropogenically (e. g., mining) significantly influenced areas such as Saarland, Saxony, the Ruhr area, etc., but also as an alternative to seismics with a comparably high value. The proposed multiresolution methods provide 3D separation of local geologic features by mollifier wavelet decomposition. Moreover, as decorrelation of the gravimetric and/or magnetometric signatures, geologic formation transitions can be increasingly prepared from sufficiently dense and accurate data systems.

From a scientific and conceptual point of view, decorrelative exploration indeed offers a new component of future geoexplorative application that should not be underestimated. On 08/18/2022, the Bundesamt für Wirtschaft und Ausführkontrolle published the "Richtlinie für die Bundesförderung für effiziente Wärmenetze (BEW)" (Guideline for Federal Funding for Efficient Heat Networks) and on 09/26/2022, the associated leaflets. The aim is to significantly increase the share of renewable energies and waste heat in heating networks in Germany. Existing district heating networks are to be transformed to renewable energy and waste heat by 2050. At least a 75% share of renewable energy and waste heat is to be planned in the construction of new district heating networks. Particular attention is paid to deep geothermal energy, the complete prospecting and exploration of which is subsidized with 50% (up to a maximum of 2 million €). For the investment in wells, plants and district heating network, a further 100 million € (maximum 40%) is available.

Without doubt we are allowed to give the following conclusions:

- Geothermal energy is an important building block in the increasing demand for renewable energy production. Deep geothermal energy means a secured and stable basic supply of electricity and heat (cf. e. g. the explanations in [22]). The advantages, such as seasonal independence from weather phenomena, local production, environmentally friendly character, visually inconspicuous and low land requirements, optimal energy utilization through cogeneration, are contrasted as a critical moment by the reduction of the discovery risk, the management of which can only succeed on the basis of increased research activity with geomathematics as a key technology.

- In general, gravimetric potentials and – somewhat attenuated in signature drawing – magnetometric potentials are also very smooth (see, e. g. [45, 46] for an overview of methods in common geophysical use), so that the potentials themselves open up exploration statements only by applying decorrelating wavelet methods. This is the reason why until today, e. g. in the geothermal scene, potential methods are largely considered applicable only in connection with sharply occurring geologic formation boundaries. However, our claim with this contribution is to document that in the future also weaker anomalies can be detected and the geologic structure of an area can be modeled interpretably by gravimetry and magnetometry and their synoptic possibilities.

- Further economization should be sought with respect to the cost and effort of gravimetric (and here in particular GNSS-based tachymetric) measurement methods to be considered indispensably more advantageous than seismic. Finally, the social compatibility of gravimetry and magnetometry compared to reflection seismics should be noted positively for investors, offices, and especially for the affected population.

- This work, whose numerical results are largely obtained on the test example of the Saarland, pursues, using the research contents from the projects GEOFÜND (PtJ FKZ 0325512A), SPE (PtJ FKZ 0324061), and SYSEXPL (PtJ FKZ 03EE4002A), the sustainable goal to show that an improvement of the cost/risk ratio is achievable by applying appropriate geomathematical solution methods under the condition of an adequate data situation. The special starting point for the integration of potential methods as an important mainstay in exploration is the claim to enable a reduction of the project risk with low financial expenditure.

Nevertheless, a considerable residual risk is canonically unavoidable for the future as well, independent of even the most conscientious exploration investigations (cf., e. g., [72, 73]). The reason lies, as already mentioned, in the specific character of gravitational and/or magnetic potential methods as "inverse problems" over a local exploration area. These are mathematically describable by a Fredholm integral equation of the first kind (cf. [54] for more details), which forces to infer from the *effect* (i. g., the potential) – primarily in points of the Earth's surface – the *cause* (i. e., the contrast function) describing the geology in the Earth's interior. Here, the "source kernel function" (often called Green's function) resulting from the exploration method used represents a singular integral kernel. The circumstance of "ill-posedness" of the integral equation (cf., e. g., [34, 64, 73]) implies a perpetual intrinsically existing limiting uncertainty for a geothermal project. From the investor's point of view, this is undoubtedly the biggest challenge that cannot be overcome without geomathematics as a synoptic key technology.

- Our goal succeeds through the key ideas and concepts of decorrelative gravimetry and magnetometry through specific realistic mollifier wavelets and their synoptic perspectives. Indeed, the GEOFÜND (PtJ FKZ 0325512A), SPE (PtJ FKZ 0324061), and SYSEXPL (PtJ FKZ 03EE4002A) projects provide evidence that the development of increasingly powerful gravimeters and magnetometers with significantly improved measurement accuracy will allow not only incisive anomalies (such as that of a plume) but also fainter signatures to be detected and modeled in the future. Future decorrelative gravimetry and magnetometry are thus new exploration techniques that result as a canonical innovation from the combination of novel measurement and modeling techniques.

Altogether, it is our belief that a geoscientific consortium (e. g., a team as listed in the beginning of this book) including geomathematics as a key technology fulfils the qualifications in contributing significantly to the more appropriate answers for all types of geothermal energy systems, i. e., deep as well as near surface geothermal systems.

Finally the following hint should be allowed: The "Federal Ministry for Economic Affairs and Climate Action, Berlin" (Bundesministerium für Wirtschaft und Klimaschutz (BMWK)) published the "Eckpunkte für eine Erdwärmekampagne - Geothermie in der Wärmewende" (cornerstones for a geothermal campaign - geothermal energy in the heat transition) on 11/09/2022. By 2030, 100 additional geothermal projects are to provide heat. Given the existing licensed capacity in Central Europe to conduct 3D seismic surveys, typically covering areas of 50 - 150 km^2 and with pure survey times of several weeks to months, this target is utopian. However, if the decorrelative exploration methods described here are used intensively, it might be conceivable to achieve the desired goal.

9.2 Outlook

As in almost all other exploratory methodologies known until now, a non-underestimated difficulty arising in mollifier decorrelation methods presented so far is that the solution of linear systems of equations is unavoidable. This requirement must definitely be critically questioned with respect to data storage and, in particular, a stable solution technique. So, a new field of research is of tremendous interest, namely a methodology for solving physics-driven inverse problems, such as gravimetry and magnetometry, based on the concept that unknown field sources should be recovered by a suitable sampling strategy using weighted sums in terms of exponentials. In doing so we are indeed able to circumvent linear systems, as far as the data are given on uniform grids. In other words, the price we have to pay are uniformly prescribed input datasets, i. e., potential data on lattice structures. Therefore we have to do measurements on lattice structures or we have to use mathematical interpolatory/extrapolatory procedures to generate an artificial data system on lattices in Euclidean space $\mathbb{R}^3$.

As a matter of fact, some approaches have been put forward for recovering the unknown frequencies and amplitudes of sums in terms of exponentials (see, e. g., [19, 20, 67, 68, 90, 129], and the references therein) usually via Prony's method or a proper variant of it. Concerning algorithms based on Prony's method we are led to e. g. [4, 144, 148]. In fact, the use of weighted sums of exponentials with purely imaginary exponents for solving inverse source problems of diffusion, wave, and potential fields from uniform data systems is for two main reasons:

- The first reason is to generate a system of superimposed sinusoids (see also [179]), which can be efficiently solved by using multi-dimensional extensions of Prony's frequency estimation methods.
- The second reason is because the use of imaginary exponentials improves the stability of the estimation problem, since in this case the magnitude of the terms in the generalized sequence remain bounded. Explicit formulas are available for the sequence of weights, that map sensor samples to a sequence of so-called generalized measurements, thereby accepting the requirement in case of uniform data that the fundamental solution (Green's

function) to the underlying differential operator associated to the inverse problem has to satisfy generalized Strang-Fix conditions (see [168]).

As a consequence, in our present work, we are not only able to take advantage from the local support of the trial functions approximating fundamental solutions (Green's functions to physically relevant differential operators), but also from an exponentially accelerated cubature thereby providing inherently the inversion process. Unfortunately, all fundamental solutions in physics-driven inverse problems of interest (in our case, monopoles in gravimetry and dipoles in magnetometry) do not show a compact local support, so that theoretical as well as practical obligations of exponential reduction cannot be guaranteed in a straightforward manner. This is the reason why this book proposes certain mollifier regularizations of the fundamental solution, so that difference mollifier wavelets indeed represent locally supported functions. The idea of a multiscale mollifier sampling methodology suitable for decorrelation purposes of functions (signals) had been already pointed out in [68], but its efficient realization using weighted sums of exponentials with purely imaginary exponents just by cubature in the framework of inverse Fourier transform is a new field of research.

The concept presented below is essentially motivated by the work proposed by J. Murray-Bruce, P.L. Dragotti [129] and is documented in [68]. While [129] essentially deals with sensor network strategies for solving inverse source problems for diffusion and wave fields, our special interest here is in the Earth's gravitational and magnetic potential field. We demonstrate that the unknown signature of the contrast function distribution inside an area $\mathcal{B}$ of the Earth can be recovered from a sequence of so-called generalized measurements by using frequency estimation techniques for the exponential ansatz. More concretely, the weights of the sums correspond to certain expressions, which reproduce multi-dimensional exponentials. Explicit formulas are derived for the sequence of weights, that map gravimetric/magnetometric samples to the sequence of generalized measurements, thereby observing the fact in case of uniform data that the fundamental solution of the Laplace differential operator satisfies the Strang-Fix condition (see [168]).

Inverse Gravimetry Sampling Decorrelation Problem. As already known, the point of departure in the case of inverse gravimetry is Poisson's differential equation, which is using the Laplace operator Δ, so that $\rho = -\Delta V$ in the region $\mathcal{B}$, where according to the Newtonian theory, V is the gravitational geopotential to the (sufficiently smooth) density distribution F. The general solution of the Poisson equation can be calculated with the help of the fundamental solution (Newtonian monopole kernel) of the Laplace operator Δ in Euclidean space $\mathbb{R}^3$

$$G(x,y) = \frac{1}{4\pi} \frac{1}{|x-y|}, \quad x, y \in \mathbb{R}^3,\ x \neq y. \tag{9.1}$$

by the improper (volumetric) Newtonian integral representation

$$V(x) = \int_{\mathcal{B}} G(x,y)\ \rho(y)\ dV(y), \quad x \in \mathbb{R}^3. \tag{9.2}$$

Moreover, in the mollifier nomenclature presented in Chapter 4, V can be substituted by V_{τ_J} given by the proper integral

$$V_{\tau_J}(x) = \int_{\mathcal{B}} G_{\tau_J}(x,y)\ \rho(y)\ dV(y) \tag{9.3}$$

with J sufficiently large and G_{τ_J} the mollifier variant of G.

So, inverse gravimetry (IG) amounts to the determination of the density distribution

$$\rho_{\tau_J}(x) = \int_{\mathcal{B}} H_{\tau_J}(|x-y|)\ \rho(y)\ dV(y), \quad x \in \mathcal{B}, \tag{9.4}$$

using the integral equation (9.3) corresponding to the gravitational potential V_{τ_J}.

In the generalized lattice sampling paradigm (see, e. g., [3, 179] and the references therein) the primary goal is to reconstruct some functions of a continuous variable from a discrete set of measurements collected on a lattice (i. e., uniform grid). Our sampling approach below is based on canonical assumptions on today's gravimetry: Suppose that a gravimeter collects a uniform sample of values of the (mollifier approximation of the) Earth's gravitational potential V_{τ_J}. Furthermore, we assume that a basic version of the Earth's gravitational potential is known, e. g., from an available Earth's gravitational model EGM. Mathematically, we interpret the a priori known information from the EGM-model in mollifier nomenclature as lowpass filter $V_{\tau_{J_0}}$ for some scale $J_0 < J$.

The inverse gravimetry sampling problem to be solved is as follows: Given the potential difference

$$(WV)_{J_0}^{J} = V_{\tau_J} - V_{\tau_{J_0}},\ J_0 < J, \tag{9.5}$$

for a (pointset of a) lattice $\Lambda \subset \mathbb{R}^3$, we intend to estimate the unknown density signature ρ_{τ_J} in $\mathcal{B}$.

The reason for using mollifier bandpass structured samples $(WV)_{J_0}^{J}$ is that the associated Newtonian mollifier bandpass kernel

$$(WG)_{J_0}^{J} = G_{\tau_J} - G_{\tau_{J_0}}, \quad J_0 < J, \tag{9.6}$$

shows a local support (for further details the reader is referred to Chapter 4).

Context of Exponential Reproduction from Uniform Translates. The main focus of our considerations will be on the following mollifier parametrization for the required density distribution

$$\rho(x) = \sum_{l=1}^{M} c_l \; \rho_{\tau_J}(|x - x_l|), \quad x \in \overline{\mathcal{B}}, \tag{9.7}$$

where $c_l \in \mathbb{R}$ is the intensity at the location $x_l \in \mathcal{B}$. For the sufficiently large scale J the kernel H_{τ_J} takes the role of the Dirac delta function. As a result of the parametrization (9.7), inverse gravimetry becomes one of estimating the unknown source parameters $\{c_l, x_l\}_{l=1,\ldots,M}$, in the spirit of [129], but now in our mollifier nomenclature of Chapter 4, by the *framework of generalized measurements* by means of the Fourier transform

$$(\rho_{\tau_J})^{\wedge}_{\mathcal{B}}(x) = \int_{\mathcal{B}} \rho_{\tau_J}(y) \; e^{-2\pi i \; x \cdot y} \; dV(y) \approx \sum_{l=1}^{M} c_l \; e^{-2\pi i \; x \cdot x_l}, \; x \in \mathbb{R}^3. \tag{9.8}$$

Obviously, we propose the use of exponentials with purely imaginary exponents as the proper choice for our sensing functions. Once again, the reasons are twofold: The first is that the left side of (9.8) becomes a system of superimposed sinusoids, which can be efficiently solved by using multi-dimensional extensions of Prony's frequency estimation methods. The second reason is because the use of imaginary exponentials improves the stability of the estimation problem, since in this case the magnitude of the terms in the generalized sequence remain bounded.

In order to compute the desired sequence of generalized measurements we again follow [129], i. e., we start from a weighted linear combination of the form

$$\sum_{g \in \Lambda} \sigma_k(g) \;\; (WG)^J_{J_0}(g, x) = e^{-2\pi i \; k \cdot x}, \quad x \in \mathbb{R}^3, \; k \in \Gamma \subset \mathbb{N}^q. \tag{9.9}$$

Then, in connection with the Newtonian mollifier theory (cf. (9.3)), it follows that

$$
\begin{aligned}
(\rho_{\mathcal{T}_J})^{\wedge}_{\mathcal{B}}(k) &= \int_{\mathcal{B}} \rho_{\mathcal{T}_J}(y)\; e^{-2\pi i\; k\cdot y}\; dV(y) \\
&= \int_{\mathcal{B}} \rho_{\mathcal{T}_J}(y) \sum_{g\in\Lambda} \sigma_k(g)\; (WG)^J_{J_0}(g,y)\; dV(y) \\
&= \sum_{g\in\Lambda} \sigma_k(g) \int_{\mathcal{B}} \rho_{\mathcal{T}_J}(y)\; (WG)^J_{J_0}(g,y)\; dV(y) \\
&= \sum_{g\in\Lambda} \sigma_k(g)\; (WV)^J_{J_0}(g), \quad k\in\Gamma,
\end{aligned} \tag{9.10}
$$

where Λ is a lattice in $\mathbb{R}^3$ and Γ is chosen as a Prony-like point system (see, e.g., the references in [68], [129]).

Thus, computing the set of generalized measurements $(\rho_{\mathcal{T}_J})^{\wedge}_{\mathcal{B}}(k)$ by taking $\sigma_k(g)$-weighted linear combinations of the gravimetric data $(WV)^J_{J_0}(g)$ is equivalent to reproducing exponentials $e^{-2\pi i\; k\cdot x}$ from space translates of the Newton mollifier wavelet kernel (3D mollifier wavelet monopole). Altogether, we are confronted with an *exponential reproduction problem* (9.9), where the only missing piece is how to obtain the coefficients $\sigma_k(g)$. Surprisingly, these coefficients are known from the literature (see [168]), because, the Newton mollifier wavelet kernel (9.6) is locally supported. Therefore, the equations (9.9) satisfy the Strang-Fix conditions (cf. [168]), so that we are immediately able to deduce that

$$
\sigma_k(g) = \frac{e^{-2\pi i k\cdot g}}{\left((WG)^J_{J_0}\right)^{\wedge}_{\mathbb{R}^3}(k)}, \qquad k\in\Gamma,\; g\in\Lambda. \tag{9.11}
$$

Estimating the Parameters of Superimposed Exponentials. With the access to the desired coefficients, $\sigma_k(g)$, that are capable of reproducing exponentials from the translates of the Newtonian wavelet monopole kernel, all we have to do at the end is the evaluation of $\{(\rho_{\mathcal{T}_J})^{\wedge}_{\mathcal{B}}(k)\}_{k\in\Gamma}$, to extract the unknown source parameters $\{c_l, x_l\}_{l=1,\ldots,M}$.

$$
\begin{aligned}
(\rho_{\mathcal{T}_J})^{\wedge}_{\mathcal{B}}(k) &\approx \sum_{l=1}^{M} c_l\; e^{-2\pi i\; k\cdot x_l} \\
&= \sum_{g\in\Lambda} \frac{e^{-2\pi i k\cdot g}}{\left((WG)^J_{J_0}\right)^{\wedge}_{\mathbb{R}^3}(k)}\; (WV)^J_{J_0}(g), \qquad k\in\Gamma.
\end{aligned} \tag{9.12}
$$

In fact, (9.8) forms a superposition of damped complex sinusoids. Thus, we are confronted with a Prony-like system. This system is prevalent, for example, in spectral estimation and array processing applications. Many approaches have been put forward for recovering the unknown frequencies and amplitudes, particularly for the two-dimensional case (see, e. g., [101, 118], [149]). Solving the Prony-like system allows to recover the unknown source parameters by utilizing, e. g., an ESPRIT algorithm by S. Sahnoun et al. [152], which represents an update of the two-dimensional ESPRIT algorithm of [149]. Specifically, in the framework as presented here, an adapted ESPRIT algorithm can be found in [129].

Non-Uniform distribution of Gravimeter Data. A simple, yet effective scheme for handling a non-uniform distribution of gravimeter data is to interpolate the field and resample it on a uniform grid. In so doing, we can obtain an approach that can still exploit the closed form expression as above even when the original data points are irregularly positioned. Besides avoiding the inversion of poorly-conditioned matrices occurring in linear systems for non-uniform gravimeter data, the interpolation approach is less intensive computationally compared to matrix inversion, particularly when the coefficient matrix of the linear system is large. In experimental results, 1D splines may be used as the interpolatory family.

Finally, it should be mentioned that sampling of inverse magnetometry (IM) can be handled analogously to the methodology for inverse gravimetry (IG).

Appendix: List of Acronyms

BGI	Bureau Gravimetrique International
BKG	Bundesamt für Kartographie und Geodäsie
BMWi	Bundesministerium für Wirtschaft und Energie
BMWK	Bundesministerium für Wirtschaft und Klimaschutz
BP	official styling BP p.l.c., formerly The British Petroleum Company plc, London
CBM	Gesellschaft für Consulting, Business und Management mbH, Bexbach
CG	Gravimeter of the company Scintrex
DMT	DMT GmbH & Co. KG, Essen; vormals Deutsche Montan Technologie GmbH
EAGE	European Association of Geoscientists and Engineers
EGM	Earth's Gravitational Model
FIS-GP	Geophysics Information System
FKZ	Förderkennzeichen
GEOFÜND	Charakterisierung und Weiterentwicklung integrativer Untersuchungsmethoden zur Quantifizierung des Fündigkeitsrisikos (GEOFÜND), research project, FKZ of the PtJ 0325512
G.E.O.S.	Ingenieurgesellschaft mbH, Halsbrücke
GOCE	Gravity Field and Steady-State Ocean Circulation
GNSS	Global Navigation Satellite System
GPS	(US) Global Positioning System
GÜK	Geologische Übersichtskarte
GSM-19	Gerät der Firma Bartington zur Messung der Totalintensität des Magnetfeldes
ITWM	Fraunhofer-Institut für Techno- und Wirtschaftsmathematik, Kaiserslautern
LIAG	Leibniz-Institut für Angewandte Geophysik, Hannover
LVGL	Landesamt für Vermessung Geoinformation und Landentwicklung, Saarbrücken
LVermGeo	Landesamt für Vermessung und Geobasisinformation, Koblenz
OST	Ostschweizer Fachhochschule, Buchs
PtJ	Projektträger und Forschungszentrum Jülich
REX	Reliable Prospection and Exploration, Forschungsvorhaben der OST, Institut für Computational Engineering, Buchs, Schweiz
SPE	Satellitengestützte Potenzialverfahren zur geothermischen Exploration, Forschungsvorhaben, FKZ des PtJ 0324061
SYSEXPL	Systematische Exploration mittels (geo)magnetischer Potenzialfeldmethoden in der Geothermie, Forschungsvorhaben, FKZ des PtJ 03EE4002A/B
MS	Magnetic Susceptibility

W. Freeden, H. Nutz, *Exploratory Potential Methods in Geothermal Power Generation*,
Lecture Notes in Geosystems Mathematics and Computing,
https://doi.org/10.1007/978-3-031-54412-5

References

1. Anderle, H.J., The evolution of the South Hunsrück and Taunus borderzone. Tectonophysics, (137):101–114, 1987.
2. Andres, G., Fränkische Alb und Malmkarst des Molassebeckens. In: Grundwassergleichenkarte von Bayern 1:500 000 mit Erläuterungen. Schriftenreihe Bayer. Landesamt für Wasserwirtschaft, München, (20):23–25, 1985.
3. Aldroubi A., Gröchenig, K., Nonuniform asampling and reconstruction in shift-invariant spaces. SIAM Review, (43):585–620, 2001.
4. Aldroubi, A., Krishtal, I., Krylov subspace methods in dynamical sampling. Sampl. Theory Signal Image Process., (15):9–20, 2016.
5. Anger, G., A characterization of the inverse gravimetric source problem through extremal measures. Rev. Geophys. Space Phys., (19):299–306, 1981.
6. Aretz, A., Aufschlussanalogstudie zur geothermischen Reservoircharakterisierung des Permokarbons im nördlichen Oberrheingraben. Ph.-D. Thesis, Technischen Universität Darmstadt, 2016.
7. Arndt, D., Geologische Strukturmodellierung von Hessen zur Bestimmung von Geo-Potenzialen. Ph.-D. Thesis, Technische Universität Darmstadt, 2012.
8. Augustin, M.A., On the role of poroelasticity for modeling of stress fields in geothermal reservoirs. GEM Int. J. Geomath., (3):67–93, 2012.
9. Augustin, M.A., A method of fundamental solutions in poroelasticity to model the stress field in geothermal reservoirs. Ph.-D. Thesis, University of Kaiserslautern, Geomathematics Group, 2014.
10. Augustin, M.A., A method of fundamental solutions in poroelasticity to model the stress field in geothermal reservoirs. Lecture Notes in Geosystems Mathematics and Computing, Bd. 1: Springer International Publishing AG, Birkhäuser Verlag, 2015.
11. Augustin, M., Freeden, W., Nutz, H., About the importance of the Runge–Walsh concept for gravitational field determination. In: Handbook of Mathematical Geodesy, W. Freeden, M.Z. Nashed (Eds.), Geosystems Mathematics, Springer International Publishing, Basel, New York, Heidelberg, 2018.
12. Axelsson, G., Gunnlaugsson, E., Long-term monitoring of high- and lowenthalpy fields under exploitation. In: World Geothermal Congress 2000, Pre- Congress Course, Kokonoe, Japan, 2000.
13. Backus, G.E., Parker, R., Constable, C., Foundations of geomagnetism. Cambridge University Press, Cambridge, 1996.
14. Backus, G.E., Gilbert, F., Numerical applications of a formalism for geophysical inverse problems. Geophys. J.R. Astron. Soc., (13):247–276, 1967.
15. Backus, G.E., Gilbert, F., The resolving power of gross Earth data. Geophys. J. R. Astron. Soc., (16):169-205, 1968.
16. Backus, G.E., Gilbert, F., Uniqueness in the inversion of inaccurate gross Earth data. Phil. Transact. R. Soc. London A, (266):123–192, 1970.
17. Bartington Instruments Limited: Operation manual for Grad-13 digital three-axis magnetic gradiometer.

W. Freeden, H. Nutz, *Exploratory Potential Methods in Geothermal Power Generation*, Lecture Notes in Geosystems Mathematics and Computing,
https://doi.org/10.1007/978-3-031-54412-5

18. Baumeister, J., Stable solution of inverse problems. Vieweg Braunschweig, 1987.
19. Beylkin, G., Monzón, L., On Approximation of functions by exponential sums. Appl. Comput. Harm. Analysis, (19):17–48, 2005.
20. Beylkin, G., Monzón, L., Approximation of functions by exponential sums revisited. Appl. Comput. Harm. Analysis, (28):131–149, 2010.
21. Bauer, M., Freeden, W., Jacobi, H., Neu, T., Energiewirtschaft 2014 (Fakten und Chancen der Tiefen Geothermie). Springer Spektrum, Springer Fachmedien Wiesbaden, 2014.
22. Bauer, M., Freeden, W., Jacobi, H., Neu, T. (Hrsg.), Handbuch Tiefe Geothermie. Springer Spektrum, Heidelberg, 2014.
23. Bauer, M., Freeden, W., Jacobi, H., Neu, T. (Hrsg.), Handbuch Oberflächennahe Geothermie. Springer Spektrum, Heidelberg, 2018.
24. Berg, G., Blick, C., Cieslack, M., Freeden, W., Hauler, Z., Nutz, H., Gravimetric measurements, gravity anomalies, geoid, quasigeoid: Theoretical background and multi-scale modeling. In: Handbuch der Geodäsie, Mathematische Geodäsie/Mathematical Geodesy, W. Freeden (Bd. Hrsg.), W. Freeden, R. Rummel (Hrsg.), Springer Spektrum, Heidelberg, (2):1117–1182, 2020.
25. Billette, F.J., Brandsberg-Dahl, S., The 2004 BP velocity benchmark: 67th Annual EAGE Meeting, EAGE, Expanded Abstracts, Soc. Expl. Geophys., B305, 2004.
26. Birner, J., Fritzer, T., Jodocy, M., Savvatis, A., Schneider, M., Stober, I., Hydraulische Eigenschaften des Malmaquifers im Süddeutschen Molassebecken und ihre Bedeutung für die geothermische Erschließung. Z. geologische Wissenschaften. (40):133–156, 2012.
27. Blakely, R.J., Potential theory in gravity and magnetic application. Cambridge University Press, Cambridge, 1996.
28. Blick, C., Multiscale potential methods in geothermal research: Decorrelation reflected post-processing and locally based inversion. Ph.-D. Thesis, University of Kaiserslautern, Geomathematics Group, 2015.
29. Blick, C., Eberle, S., Multiscale density decorrelation by Cauchy-Navier wavelets. GEM Int. J. Geomath., (10):1–31, 2019.
30. Blick, C., Eberle, S., A survey on multiscale mollifier decorrelation of seismic data. GEM Int. J. Geomath., (12):16, 2021.
31. Blick, C., Freeden, W., Nutz. H., Feature extraction of geological signatures by multiscale gravimetry. GEM Int. J. Geomath, (8):57–83, 2017.
32. Blick, C., Freeden, W., Nutz, H., Innovative Explorationsmethoden in der Gravimetrie und der Reflektionsseismik. In: Handbuch Oberflächennahe Geothermie, M. Bauer, W. Freeden, H. Jacobi, T. Neu (Hrsg.), Springer, Berlin, 221–256, 2018.
33. Blick, C., Freeden, W., Nutz, H.: Gravimetry and exploration. In: Handbook of Mathematical Geodesy, W. Freeden, M.Z. Nashed (Hrsg.), Geosystems Mathematics, Birkhäuser, Springer International, 687–751, 2018.
34. Blick, C., Freeden, W., Nashed, M.Z., Nutz, H., Schreiner, M., Inverse magnetometry: Mollifier magnetization distribution from geomagnetic field data, Lecture Notes in Geosystems Mathematics and Computing, Birkhäuser, Springer Nature, Cham, Switzerland, 2021.
35. Bödvarsson, G., Physical characteristics of natural heat sources in Iceland. In: Proc. UN Conf. on New Sources of Energy, volume 2. United Nations, 1964.
36. Dachroth, W., Zur stratigraphischen Korrelation auf Grund des gesteinsmagnetischen Verhaltens und Anwendung auf das Oberrotliegende und den Buntsandstein von Rheinland-Pfalz und Saar, Ph.-D. Thesis, Saarbrücken, 1969.
37. Dachroth, W., Erdmagnetische Messungen im östlichen Saarland. Erdöl und Kohle, 401–404; Hamburg, 1970.
38. Dallmeyer, R.D., Franke, W., Weber, K., Pre-permian geology of central and eastern Europe, Springer, Berlin, 579–595, 1995.
39. Dorsch, K., Pletl, C., Bayerisches Molassebecken - Erfolgsregion der Tiefengeothermie in Mitteleuropa. Geothermische Energie, (73):14–18, 2012.

40. Dumas, P., Landolina, S., Strategic research priorities for geothermal technology. European Technology Platform on Renewable Heating and Cooling, 2012.
41. Engel, H., Zur Tektogenese des Saarbrücker Hauptsattels und der Südlichen Randüberschiebung. In: Beiträge zur Tiefentektonik westdeutscher Steinkohlenlagerstätten, G. Drozdzewski, H. Engel, R. Wolf, V. Wrede (Eds.), Geol. L.-Amt Nordrh.-Westf., 217–235, 1985.
42. Engl, H., Integralgleichungen. Springer, Berlin, 1997.
43. Engl, H. W., Hanke, M., Neubauer, A., Regularization of inverse problems. Kluwer, Dordrecht, 1996.
44. Engl, H., Louis, A.K., Rundell, W., (Eds.), Inverse problems in geophysical applications. SIAM, Philadelphia, 1997.
45. Ernstson, K., Geologische und geophysikalische Untersuchungen. In: Handbuch Tiefe Geothermie, M. Bauer, W. Freeden, H. Jacobi, T. Neu (Eds.), Springer Spektrum, Heidelberg, 19–80, 2014.
46. Ernstson, K., Alt, W., Gravimetrie und Geomagmetik bei Geothermieprojekten: Verständnis und Missverständnis. World of Mining, (65):115–122, 2013.
47. Falke, H., Kneuper, G., Das Karbon in limnischer Entwicklung. SeptiÃĺme CongrÃĺs International de Stratigraphie et de GÃľologie du CarbonifÃĺre, (1):49–67, Krefeld, 1972.
48. Franke, R.D., Franke, W., Weber, K., Pre-permian geology of central and eastern Europe. 579–595, Springer, Berlin, 1995.
49. Franke, W., The mid-European segment of the Variscides: tectonostratigraphic units, terrane boundaries and plate tectonic evolution. In: Orogenic Processes: Quantification and Modelling in the Variscan Belt, W. Franke, V. Haak, O. Oncken, D. Tanner (Eds.), Geological Society Special Publications, London, (179):35–61, 2000.
50. Freeden, W., On the approximation of external gravitational potential with closed systems of (trial) functions. Bull. Geod., (54):1–20, 1980.
51. Freeden, W., On approximation by harmonic splines. Manuscr. Geod., (6):193–244, 1981.
52. Freeden, W., Multiscale modeling of spaceborne geodata. Teubner, Stuttgart, 1999.
53. Freeden, W., Geomathematics: Its role, its aim, and its potential. In: Handbook of Geomathematics, W. Freeden, M.Z. Nashed, T. Sonar, (Hrsg.), Springer, Heidelberg, Vol. 1, 2nd ed., 3–78, 2015.
54. Freeden, W., Decorrelative mollifier gravimetry – basics, ideas, concepts, and examples. Geosystems Mathematics, Springer Nature, Birkhäuser, Cham, Switzerland, 2021.
55. Freeden, W., Bauer, M., Dekorrelative Gravimetrie – Ein innovativer Zugang in Exploration und Geowissenschaften. Springer Spektrum, Berlin, 2020.
56. Freeden, W., Blick, C., Signal decorrelation by means of multiscale methods. World of Mining, (65):304–317, 2013.
57. Freeden W., Gerhards C., Geomathematically oriented potential theory. Chapman & Hall/CRC Press, Boca Raton, 2013.
58. Freeden, W., Gutting, M., Integration and cubature methods. Chapman and Hall, CRC Press, Boca Raton, London, New York, 2018.
59. Freeden, W., Heine, C., Geothermie – Ein kurzer Abriss. In: Handbuch Oberflächennahe Geothermie, M. Bauer, W. Freeden, H. Jacobi, T. Neu (Eds.), Springer, Heidelberg, 1–22, 2018.
60. Freeden, W., Heine, C., Nashed M.Z., An invitation to Geomathematics. Lecture Notes in Geosystem Mathematics and Computing, 2019.
61. Freeden, W., Michel, V., Multiscale potential theory (with applications to Geoscience). Birkhäuser, Boston, 2004.
62. Freeden, W., Nashed, M.Z., Ill-posed problems: Operator methodologies of resolution and regularization approaches. In: Handbook of Mathematical Geodesy, Geosystems Mathematics, Freeden, W., Nashed, M.Z. (Eds.), Springer International Publishing, Birkhäuser, Basel, New York, Heidelberg, Geosystems Mathematics, 201-314, 2018a.

63. Freeden, W., Nashed, M.Z., Inverse gravimetry as an ill-posed problem in Mathematical Geodesy. In: Handbook of Mathematical Geodesy. Geosystems Mathematics, W. Freeden, M.Z. Nashed (Eds.), Springer, International Publishing, Birkhäuser, Basel, New York, Heidelberg, Geosystems Mathematics, 641–685, 2018b.
64. Freeden, W., Nashed, M.Z., Inverse gravimetry: Background material and multiscale mollifier approaches. GEM Int. J. Geomath., (9):199–264, 2018c.
65. Freeden, W., Nashed, M.Z., Operator-theoretic and regularization approaches to ill-posed problems. GEM Int. J. Geomath., (9):1–115, 2018d.
66. Freeden, W., Nashed, M.Z. (Eds.), Handbook of Mathematical Geodesy, Springer International Publishing, Birkhäuser, Basel, New York, Heidelberg, Geosystems Mathematics, 2018g.
67. Freeden, W., Nashed, M.Z., Inverse gravimetry: Density signatures from gravitational potential data. In: Handbuch der Geodäsie, Mathematische Geodäsie/Mathematical Geodesy, W. Freeden (Band Ed.), W. Freeden, R. Rummel (Eds.), Springer Spektrum, Heidelberg, 969-1052, 2020.
68. Freeden, W., Nashed, M.Z., Recovery methodologies: Regularization and sampling. American Mathematical Society, Applied Mathematics, Mathematical Surveys and Monographs (SURV), Volume 274, Providence, R.I., 2023.
69. Freeden, W., Nashed, M.Z., Schreiner, M., Spherical sampling. Geosystems Mathematics, Springer International Publishing, Basel, New York, Heidelberg, 2018.
70. Freeden, W., Nutz, H., Satellite gravity gradiometry as tensorial inverse problem. GEM Int. J. Geomath., (2):177–218, 2011.
71. Freeden, W., Nutz, H., Mathematik als Schlüsseltechnologie zum Verständnis des Systems "Tiefe Geothermie". Jahresber. Deutsch. Math. Vereinigung (DMV), (117):45–84, 2015.
72. Freeden, W., Nutz, H., Inverse Probleme der Geodäsie: Ein Abriss mathematischer Lösungsstrategien. In: Handbuch der Geodäsie, Mathematische Geodäsie/Mathematical Geodesy, W. Freeden (Bd. Ed.), W. Freeden, R. Rummel (Eds.), Springer Spektrum, Heidelberg, 65–90, 2020.
73. Freeden, W., Nutz, H., Inverse problems in Geodesy: An outline of mathematical solution strategies. In: Encyclopedia of Geodesy, Encyclopedia of Earth Sciences Series, M.G. Sideris (Ed.), Springer Nature, Switzerland, Cham, doi.org/10.1007/978-3-319-02370-0_168-1, 2022.
74. Freeden W., Nutz H., Rummel R., Schreiner M., Satellite gravity gradiometry (SGG): Methodological foundation and geomathematical advances. In: Handbuch der Geodäsie, Mathematische Geodäsie/Mathematical Geodesy, W. Freeden (Bd. Ed.), W. Freeden, R. Rummel (Eds.), Springer Spektrum, Heidelberg, (2):1185–1256, 2020.
75. Freeden, W., Sansó, F., Interactions, acquisitions, and open problems. IX Hotine-Marussi Symposium on Mathematical Geodesy, P. Novák, M. Crespi, N. Sneeuw, F. Sansó (Eds.), Springer, 219-250, 2018.
76. Freeden, W., Schneider, F., Regularization wavelets and multiresolution. Inverse Problems, (14):493–515, 1998.
77. Freeden, W., Schreiner, M., Local multiscale modeling of geoid undulations from deflections of the vertical. J. Geodesy, (79):641–651, 2006.
78. Freeden, W., Schreiner, M., Mathematical Geodesy: Its role, its aim, and its potential. In: Handbuch der Geodäsie, Mathematische Geodäsie/Mathematical Geodesy, W. Freeden (Bd. Ed.), W. Freeden, R. Rummel (Ed.), Springer Spektrum, Heidelberg, 3–64, 2020.
79. Freeden, W., Schreiner, M., Spherical functions of mathematical geosciences: A scalar, vectorial, and tensorial setup. 1. Auflage, Advances in Geophysical and Environmental Mechanics and Mathematics, Springer, Berlin, Heidelberg, 2009, 2. Auflage, Geosystems Mathematics, Birkhäuser, Springer Nature, Cham, Switzerland, 2022.

80. Freeden, W., Witte, B., A combined (spline-)interpolation and smoothing method for the determination of the gravitational potential from heterogeneous data. Bull. Géod., (56):53–62, 1982.
81. Friedrichs, K.O., Die Identität schwacher und starker Erweiterungen von Differentialoperatoren. Transact. Amer. Math. Soc., (55):132-151, 1944.
82. GEMLink 5.4: Help and Operational Manual. GEM System Inc, 2017.
83. Gao, Q., Cheng, D., Wang, Y., Li, S., Wang, M., Yue, L., Zhao, J., Compensation method for diurnal variation in three-component magnetic survey. Appl. Sci. 10, 986, doi:10.3390/app10030986, 2020.
84. Gehringer, M., Loksha, V., Handbook on planning and financing geothermal power generation. – ESMAP (Energy Sector Management Assistence Programm), Main Findings and recommandations. The International Bank for Reconstruction and Developement, Washington, 2012.
85. Georgsson, L.S., Friedleifsson, I. B, Geothermal energy in the world from energy perspective. In: Short Course IV on Exploration for Geothermal Resources, 1–22, 2009.
86. Gerhards, C., Spherical multiscale methods in terms of locally supported wavelets: Theory and application to geomagnetic modeling. Ph.-D. Thesis, University of Kaiserslautern, Geomathematics Group, 2011.
87. Goldreich, A.H., Die Bodenbewegungen im Kohlenrevier und deren Einflüsse auf die Tagesoberfläche, Springer, Berlin, Heidelberg, 1926.
88. Grebe, H., Leppla, A. erläutert von Leppla, A., Erläuterungen zur geologischen Specialkarte von Preussen und den Thüringischen Staaten. Blatt Buhlenberg. In: Vertrieb bei der Simon Schropp'schen Hof-Landkartenhandlung (J. H. Neumann), Berlin W., 1898.
89. Groetsch, C.W., The theory of Tikhonov regularization for Fredholm equations of the first kind. Pitman, Boston, 1984.
90. Hackbusch, W., Entwicklungen nach Exponentialsummen. Technical Report. Max-Planck-Institut für Mathematik in den Naturwissenschaften, Leipzig, 2010.
91. Hackbusch, W., Khoromoskij, B.N., Klaus, A., Approximation of functions by exponential sums based on the Newton-type optimisation. Technical Report, Max-Planck-Institut für Mathematik in den Naturwissenschaften, Leipzig, 2005.
92. Hadamard, J., Lectures on the Cauchy problem in linear partial differential equations. Yale University Press, New Haven, 1923.
93. Hadamard, J., Sur les problÃĺmes aux dÃľrivÃľs partielles et leur signification physique. Princeton Univ. Bull., (13):49–52, 1902.
94. Hammons, T. J., Geothermal power generation: Global perspectives, technology, direct uses, plants, drilling and sustainability worldwide. In: Electricity Infrastructures in the Global Marketplace, pages 195–234. InTech, 2011.
95. Hedemann, H.-A., Wissmann, W., Die Gebirgstemperaturen in der Bohrung Saar 1 und ihre Beziehungen zum geologischen Bau. In: Geologisches Jahrbuch: Die Tiefbohrung Saar 1, E. Schweizerbart'sche Verlagsbuchhandlung (Nägele u. Obermiller), 91–305, Hannover, 1976.
96. Helms, L.L., Introduction to Potential Theory. Wiley-Interscience, New York, 1969.
97. Henk, A., Das Saar-Nahe-Becken. Eine geodynamische Beckenanalyse. Die Geowissenschaften, 11. Jahrg. (8):268–273, 1993.
98. Henk, A., Late orogenic basin evolution in the Variscan Internides: the Saar-Nahe Basin, southwest Germany. Tectonophysics, (223):273–290, 1993.
99. Hertle, M., Numerische Simulation der geologischen Entwicklungsgeschichte des permokarbonen Saar-Nahe-Beckens. Ph.-D. Thesis, RWTH Aachen, 2003.
100. Hinze, W., Von Frese, R., Saad, A., Applications of the gravity and magnetic methods to subsurface exploration, Chapters 1–4, Updated 4/4/2015. Resources for Gravity and Magnetic Exploration: Principles, Practices, and Applications, 415–421, Cambridge University, 2015.

101. Hua, Y., Sarkar, T.K., Matrix pencil method for estimating parameters of exponentially damped/undamped sinusoids in noise. IEEE Transactions on Acoustics, Speech, and Signal Processing, (38):814–824, 1990.
102. Ilyasov, M., A tree algorithm for Helmholtz potential wavelets on non-smooth surfaces: Theoretical background and application to seismic data postprocessing. Ph.-D. Thesis, University of Kaiserslautern, Geomathematics Group, 2011.
103. IAGA: International Geomagnetic Reference Field, 2020, 13th generation. International Association of Geomagnetism and Aeronomy, Division V, Working Group V-MOD. http://www.ngdc.noaa.gov/IAGA/vmod/igrf.html.
104. Jacobs, F., Meyer, H., Geophysik - Signale aus der Erde. Teubner Verlag, Leipzig, 1992.
105. Kammerer, W.J., Nashed, M.Z., The convergence of the conjugate gradient method for singular linear operator equations. SIAM J. Numer. Anal., (9):165–181, 1972a.
106. Kammerer, W.J., Nashed, M.Z., Iterative methods for best approximate solutions of linear integral equation of the first and second kind. J. Math. Anal. Appl., (40):547–573, 1972b.
107. Kellogg, O.D., Foundations of potential theory. Frederick Ungar Publishing Company, New York, 1929.
108. Kirsch, A., An introduction to the mathematical theory of inverse problems. Springer, Heidelberg, 1996.
109. Knödel, K., Krummerl, H., Lange, G., Handbuch zur Erkundung des Untergrundes von Deponien. Bd. 3 Geophysik, Springer-Verlag Berlin Heidelberg, 2. Aufl., 2005.
110. Korsch, R.J., Schäfer, A., Geological interpretation of DEKORP deep seismic reflection profiles 1C and 9N across the Variscan Saar-Nahe Basin, south-west Germany. Tectonophysics, (191):127–146, 1991.
111. Korsch, R.J., Schäfer, A., The Permo-Carboniferous Saar-Nahe-Basin, south-west Germany and north-east France: basin formation and deformation in a strike-slip regime. Geol. Rundschau, (84): 293–318, 1995.
112. Kossmat, F., Gliederung des varistischen Gebirgsbaues. Abhandlungen des Sächsischen Geologischen Landesamtes, (1):1–39, Leipzig, 1927.
113. Kretz, B., Multiscale mollifier technique in poroelasticity with an introduction of thermal aspects. Ph.-D. Thesis, University of Siegen, Geomathematics Group, 2022.
114. Küstner, W., K-Ar-Datierungen an detritischen Muskoviten und Sm-Nd-Modellalter prä- und synorogener schwach metamorpher Sedimente im Rhenoherzynikum Grundlegende Daten zur Quantifizierung orogener Prozesse am Beispiel der Varisziden. Ph.-D. Thesis, Georg-August-Universität zu Göttingen, 2000.
115. Leweke, S., Michel, V., Telschow, R., On the uniqueness of gravitational and magnetic field data inversion. In: Handbook of Mathematical Geodesy, Geosystems Mathematics, W. Freeden, M.Z. Nashed (Eds.), Springer International Publishing, Birkhäuser, Basel, New York, Heidelberg, 883–920, 2018.
116. Lieusternik, L. A., Sobolev, V. J., Elements of Functional Analysis. New York, 1961.
117. Louis, A. K., Inverse und schlecht gestellte Probleme. Teubner, 1989.
118. Maravi'c, I., Vetterli, M., Exact sampling results for some classes of parametric nonbandlimited 2D-signals. IEEE Transactions on Signal Processing, (52):175–189, 2004.
119. Marti, J.T., On the convergence of an algorithm computing minimum-norm solutions of ill-posed problems. Math. Comp., (34):521–527, 1980.
120. Martin, G.S., Marfurt, K.J., Larsen, S., Marmousi-2: An updated model for the investigation of AVO in structurally complex areas. In: Proceedings, Society of Exploration Geophysicists Annual Meeting, Salt Lake City, 2002.
121. Meschede, M., Geologie Deutschlands. Ein prozessorientierter Ansatz, Springer Spektrum, Berlin, Heidelberg, 2015.

122. Meyer, R. K. F., Schmidt-Kaler H., Jura. In: Bayerisches Geologisches Landesamt (Ed.), Erläuterungen zur Geologischen Karte von Bayern 1:500 000. 4. Aufl.: 90–111; München, 1996.
123. Michel, V., A multiscale method for the gravimetry problem: Theoretical and numerical aspects of harmonic and anharmonic modeling. Ph.-D. Thesis, University of Kaiserslautern, Geomathematics Group, 1999.
124. Militzer, H., Weber, F., Angewandte Geophysik, Bd. 1: Gravimetrie und Magnetik. Springer, Wien, New York, 1984.
125. Moeck, I., Vorreiter der weltweiten Entwicklung geothermischer Ressourcen: Das Bayerische Molassebecken. GTE, (81):22–26, 2015.
126. Mongillo, M., International efforts to promote global sustainable geothermal development. In: GIA Annual Report Executive Summary, 1–19, 2011.
127. Müller, C., Aspects of differential equations in Mathematical Physics. In: Partial Differential Equations and Continuum Mechanics, R. E. Langer, Ed., The University of Wisconsin Press, Madison, 3–8, 1961.
128. Morozov, V.A., Methods for solving incorrectly posed problems. Springer, New York, 1984.
129. Murray-Bruce, J., Dragotti, P.L., A sampling framework for solving physics-driven inverse source problems. IEEE Transactions on Signal Processing, (65):6365–6380, 2017.
130. Nashed, M.Z., Generalized inverse, normal solvability and iteration for singular operator equations. In: Nonlinear functional analysis and applications, 311–359. L.B. Rall (Ed.), Academic, New York, 1971.
131. Nashed, M.Z. (Ed.), Generalized inverses and applications. Academic Press, New York, 1976a.
132. Nashed, M.Z., Aspects of generalized inverses in analysis and regularization. In: Generalized Inverses and Applications. Academic Press, New York, 193-244, 1976b.
133. Nashed, M.Z., On moment-discretization and least-squares solutions of linear integration equations of the first kind. J. Math. Anal. Appl., (539:359–366, 1976c.
134. Nashed, M.Z.: Operator-theoretic and computational approaches to ill-posed problems with applications to antenna theory. IEEE Trans. Antennas Propagation, (29), 220-231, 1981.
135. Nashed, M.Z., Inverse problems, moment problems and signal processing: Un menage a trois. Mathematics in Science and Technology, A.H. Siddiqi, R.C. Singh, P. Manchanda (Eds.), World Scientific, New Jersey, 1–19, 2010.
136. Nashed, M.Z., Scherzer, O., Inverse problems, image analysis and medical imaging (contemporary mathematics). American Methematical Society 313, Providence, R.I., 2002.
137. Nashed, M.Z., Wahba, G., Regularization and approximation of linear operator equations in reproducing kernel spaces. Bull. Am. Math. Soc., (80):1213–1218, 1974a.
138. Nashed, M.Z., Wahba, G., Convergence rates of approximate least squares solutions of linear integral and operator equations of the first kind. Math. Comput., (28):69–80, 1974b.
139. Nashed, M.Z., Walter, G.G., General sampling theorems for functions in reproducing kernel Hilbert space. Math. Contr. Signals Syst., (4):363–390, 1991.
140. Nashed, M.Z., Walter, G.G., Reproducing kernel Hilbert space from sampling expansions. Contemporary Math., (190):221–226, 1995.
141. Natterer, F., The finite element method for ill-posed problems. RAIRO Analyse Numerique, (11):271–278, 1977.
142. Nettleton, L.L., Gravity and magnetics in oil prospecting. McGraw-Hill Book Company, Inc., New York, 1976.
143. Oberbergamt des Saarlandes, Risswerke der Grube Luisenthal, 2018.

144. Peter, T., Plonka, G., A generalized Prony method for reconstruction of sparse sums of eigenfunctions of linear operators. Inverse Problems, 29(2): 025001, 21 pp., 2013.
145. Pirrung, M., Forschungsprojekt FluviMag: Fluviatiler Transport von Magnetomineralen: 5.4. Bergbau an der Saar. Institut für Geowissenschaften, Friedrich-Schiller-Universität Jena, V. 6, 2019.
146. Phillips, W.S., Rutledge, J.T., House, L. S., Fehler, M. C., Induced microearthquake patterns in hydrocarbon and geothermal reservoirs: Six case studies. Pure Appl Geophys, (159):345–369, 2002.
147. RAG, Bürgerinformationsdienst (BID) der RAG Aktiengesellschaft https://geodaten.rag.de/mapapps/resources/apps/bid/index.html?lang=de, abgerufen am 15.02.2022.
148. Reise, G., Matz, G., Gröchenig, K., Distributed field reconstruction in wireless sensor networks based on hybrid shift-invariant spaces. IEEE Trans. Signal Process., (60):5426–5439, 2012.
149. Rouquette, S., Najim, M., Estimation of frequencies and damping factors by two-dimensional ESPRIT type methods. IEEE Transactions on Signal Processing, (49):237–245, 2001.
150. Saemundsson,K., Geothermal systems in global perspective. In: Short Course IV on Exploration for Geothermal Resources, Lake Naivasha, Kenya, November 2009. UNU-GTP, KenGen and GDC.
151. Saarbergwerke Aktiengesellschaft, Überblick über die Geologie des saarländischen Steinkohlengebirges. Beilage zu "Schacht und Heim", Sammelmappe Saarbergbau, Werkzeitung der Saarbergwerke Aktiengesellschaft, September, 1962.
152. Sahnoun, S., Usevich, K., Comon, P., Multi-dimensional ESPRIT: Algorithm, computations, and perturbation. Analysis. https://hal.archives-ouvertes.fr/hal-01360438, 2016.
153. Saltus, R.W., Blakely, R.J., Unique geologic insights from "non-unique" gravity and magnetic interpretation. GSA Today, (21):4–11, 2011.
154. Sanyal, S.K., Classification of geothermal systems – a possible scheme. In: Proceedings, 30th Workshop on Geothermal Reservoir Engineering, Stanford University, California, SGP-TR-176, 85–92, 2005.
155. Shure, L., Parker, R.L., Backus, G.E., Harmonic splines for geomagnetic modelling. Phys. Earth Planet. Inter., (28):215–229, 1982.
156. Shure, I., Parker, R.L., Langel, R.A., A preliminary harmonic spline model from Magsat data. Journal of Geophysical Research, (90):11505–11512, 1985.
157. Schäfer, A., Die Sedimente des Oberkarbons und Unterrotliegenden im Saar-Nahe-Becken. Mainzer Geowiss. Mitt., (15):239–365, 1986.
158. Schäfer, A., Stamm, R., Lakustrine Sedimente im Permokarbon des Saar-Nahe-Beckens. Z. dt. geol. Ges., (140):259–276, 1989.
159. Schimazek, J., Zur Klein- und Großtektonik des Saarbrücker Hauptsattels. Ph.-D. Thesis, Mathematisch-Naturwissenschaftliche Fakultät, Freie Universität Berlin, 1964.
160. Schneider, H., Sammlung Geologischer Führer, Band 84, Saarland, Gebr. Borntraeger Berlin – Stuttgart, ISBN 978-3443150532, 1991.
161. SCINTREX Limited: CG-5 Brochure. https://secureservercdn.net/198.71.233.206/6c3.ae8.myftpupload.com/wp-content/uploads/2018/11/CG-5-Brochure-R3.pdf, abgerufen am 24.03.2022.
162. SCINTREX Limited: CG-6 Brochure. https://secureservercdn.net/198.71.233.206/6c3.ae8.myftpupload.com/wp-content/uploads/2019/01/CG-6-Brochure_R5.pdf, abgerufen am 24.03.2022.
163. Sonar, T., 3000 Jahre Analysis. Springer, Berlin, Heidelberg, New York, 2011.
164. Song, M., Regularization-projection methods and finite element approximations for ill-posed linear operator equations. Ph.-D. Thesis, Univ. Michigan, 1978.

165. Stober, I., Jodocy, M., Geothermische Nutzhorizonte im westlichen Teil des Süddeutschen Molassebeckens. Z. geologische Wissenschaften, (39):161–172, 2011.
166. Stollhofen, H., Postvulkanische Rotliegend-Schwemmfächersysteme am Hunsrück-Südrand, Saar-Nahe-Becken, SW-Deutschland (Exkursion 13. April 2007). Jber. Mitt. oberrhein. Geol. Ver., N.F., (89):285–306, 2007.
167. Strand, O. N., Theory and methods related to the singular function expansion and Landweber's iteration for integral equations of the first kind. SIAM J. Numer. Anal., (11):798–825, 1974.
168. Strang, G., Fix, G., A Fourier analysis of the finite element variational method. Constructive Aspects of Functional Analysis, G. Geymonat, Ed., C.I.M.E., 793–840, 1973.
169. Surovitckii, L., Tresnak, J. P., Smirnov, A. V., Rock magnetic properties of sediments overlying the hydrocarbon-bearing niagaran pinnacle reefs near bear lake (Michigan, USA), AGU Fall Meeting Abstracts, Michigan Technological University, Geological and Mining Engineering and Sciences, Houghton, MI, United States, 2019.
170. Telford W., Geldart L.P. und Sheriff R.E.: Applied geophysics. Cambridge University Press, 2.Auflage, 1990.
171. Tikhonov, A.N., On the stability of inverse problems. Dokl. Akad. Nauk SSSR, (39):195–198, 1943.
172. Tikhonov, A.N., On the solution of incorrectly formulated problems and the regularization method. Dokl. Akad. Nauk SSSR, (151):501–504, 1963.
173. A.N. Tikhonov, A.N., Arsenin, V.Y., Solutions of Ill-Posed Problems. Wiley, New York, 1977.
174. Torge, W., Gravimetry. Walter de Gruyter, Berlin, New York, 1989.
175. Torge, W., Geodäsie. Walter de Gruyter, Berlin, New York, 2003.
176. Torge, W., Müller, J., Geodesy. 4th ed., Walter de Gruyter, Berlin, New York, 2012.
177. Valenta, J., Introduction to Geophysics – Lecture Notes. Czech Republic Development Cooperatio, Prague, 2015.
178. Veit, E., Geophysik und Bau des Untergrundes des Saarbrücker Hauptsattels, Geologisches Jahrbuch Reihe A Heft 27, Hannover, Hrsg.: Bundesanstalt für Geowissenschaften und Rohstoffe und Geologische Landesämter der Bundesrepublik Deutschland, 409 ff., 1976.
179. Vetterli, M., Kovacevic, J., Goyal, V.K.,, Foundations of signal processing. Cambridge University Press, 2014.
180. Versteeg, R., The Marmousi experience: Velocity model determination on a synthetic complex data set, The Leading Edge, (13):927–936, 1994.
181. Vogel, D., Weitmüller, W., Ergänzende Magnetikmessungen im Saarland 2008: Bericht zur technischen Durchführung und Auswertung, Projekt Anomalienkarten von Potentialfeldern, Leibniz-Institut für Angewandte Geophysik, Hannover, 2009.
182. Walter, R., Geologie von Mitteleuropa. 5. Aufl., E. Schweizerbart'sche Verlagsbuchhandlung, 1995.
183. Welding, L., GLITNIR geothermal research. In: United States Geothermal Energy Market Report, 1–37, 2007.
184. Widmer-Schnidrig, R., Duffner, P., Forbriger, T., Zürn, W., Observing the Earth's magnetic field in an underground observatory: a case study from BFO. Festschrift zur Verabschiedung von Prof. Dr. Bernhard Heck, KIT Scientific Publishing, Schriftenreihe des Studiengangs Geodäsie und Geoinformatik, 2018.
185. Xia X.G., Nashed M.Z., The Backus-Gilbert method for signals in reproducing Hilbert spaces and wavelet subspaces. Inverse Problems, (10):785–804, 1994.
186. Zimmerle, W., Petrographische Beschreibung und Deutung der erbohrten Schichten. In: Geologisches Jahrbuch: Die Tiefbohrung Saar 1, E. Schweizerbart'sche Verlagsbuchhandlung, 91–305, Hannover, 1976.

187. https://www.britannica.com/place/Rhineland-Palatinate), accessed 27 September 2023.
188. https://www.ensheim-saar.de, accessed 01 April 2022.
189. https://www.mineralienatlas.de/lexikon/index.php/Deutschland/Saarland, accessed 01 April 2022.
190. https://www.rlp.de/de/startseite/, accessed 27 September 2023.
191. https://www.saar-nostalgie.de/Saargruben.htm, accessed 27 September 2023.
192. https://www.sr.de/sr/home/nachrichten/dossiers/ende_des_bergbaus_an_der_saar/250_jahre_bergbau_an_der_saar100.html, accessed 27 September 2023.
193. https://www.sr.de/sr/home/nachrichten/dossiers/60_jahre_saarland/60jahre saarland_vision_saarstatut100.html, accessed 27 September 2023.

Index

W. Freeden, H. Nutz, *Exploratory Potential Methods in Geothermal Power Generation*,
Lecture Notes in Geosystems Mathematics and Computing,
https://doi.org/10.1007/978-3-031-54412-5

MIX
Papier aus verantwortungsvollen Quellen
Paper from responsible sources
FSC® C105338
FSC
www.fsc.org

If you have any concerns about our products,
you can contact us on
ProductSafety@springernature.com

In case Publisher is established outside the EU,
the EU authorized representative is:
Springer Nature Customer Service Center GmbH
Europaplatz 3, 69115 Heidelberg, Germany

Printed by Libri Plureos GmbH
in Hamburg, Germany